Too comple?
June 1987

The Edge of Infinity

Where the Universe Came From
and
How It Will End

Paul Davies

A TOUCHSTONE BOOK
Published by Simon and Schuster
NEW YORK

First Touchstone Edition, 1982

A Touchstone Book
Published by Simon & Schuster, Inc.
Simon & Schuster Building
Rockefeller Center
1230 Avenue of the Americas
New York, New York 10020

TOUCHSTONE and colophon are registered trademarks
of Simon & Schuster, Inc.

Manufactured in the United States of America

Originally published in England in 1981 by J.M. Dent & Sons Ltd.

10 9 8 7 6 5 4 3 2 1
10 9 8 7 6 5 4 3   Pbk.

Library of Congress Cataloging in Publication Data

Davies, P. C. W.
  The edge of infinity.

  Includes index.
  1. Naked singularities (Cosmology)
2. Black holes (Astronomy)   I. Title.
QB991.N34D38   1981   523.1   81-16743
                                           AACR2

ISBN 0-671-44063-2
ISBN 0-671-46062-5 Pbk.

Contents

Illustrations

Preface

Several years ago, when the bookshops began to fill with stories of black holes, I was dining with a group of scientists who had made some of the discoveries in that subject. They seemed somewhat taken aback that the topic of their daily research had suddenly become the focus of public attention. 'Whatever next?' someone asked. 'The naked singularity!' came a prompt reply. So I have written, with some amusement, a book about naked singularities. The singularity should not be regarded as an object or a thing, so much as a non-place where all known laws are suspended.

Any venture into the world of modern physics is bound to produce sensation. The subject of gravity, with its extraordinary notions of warped space and time, event horizons and catastrophic collapse of stars, provides endless fascination for scientist and layman alike. I have presented what are, to the best of present understanding, the accepted ideas of modern physics and cosmology. Much of the material, although based on published research papers, is nevertheless controversial. One thing is clear, though: regions of the universe exist where gravity is so overwhelming that it threatens all physical laws and structures – regions where we are brought to the very edge of infinity.

The reader need have no detailed background knowledge of either physics or mathematics to enjoy this book. Though many of the ideas described require some mind-bending imagination, I have tried to base most of the exposition on diagrams and pictures.

I should like to thank Dr N.D. Birrell and Dr D.C. Robinson for useful discussions.

Note on technical terms

Where possible, metric units are used. For astronomical distances the light year is frequently most appropriate. This is the distance travelled by light in one year (it is not a measurement of time) and is equivalent to about 9.5 million million kilometres, or about six million million miles. Light travels at 300,000 kilometres per second. Radio frequencies are occasionally mentioned and are measured in 'megahertz', written MHz, one megahertz being equivalent to one million cycles per second.

The concept of mass occurs frequently in this book. On Earth, mass is often used interchangeably with weight, but this can be deceptive. It is better to think of mass as the quantity of matter, or ponderousness. In some cases mass is also used as a measure of gravitating power. When 'massive bodies' are mentioned, this does not imply that they are large in size, but large in mass. Thus a neutron star is only a few kilometres in diameter, but far more massive than the Earth, which contains much less material and exerts a weaker gravitational pull.

Throughout, the word 'billion' refers to the US billion, i.e. one thousand million.

1 The cosmic connection

When public disillusionment with science began to grow during the late 1960s and early 1970s, one subject remained unscathed. Astronomy, the very first science, remains today as potent a symbol as ever for the wider perspective of man. As the heavens are scrutinized by ever more sophisticated hardware, a seemingly endless sequence of astounding new discoveries is reported in the popular press. Quasars, pulsars, neutron stars, black holes . . . weird-sounding objects with enigmatic and often bizarre properties, continue to excite the imagination of scientist and layman alike.

The compulsive fascination to know what is going on out there in the universe is by no means a recent fad. It reflects a deep-rooted cultural awareness of the cosmic connection that has its origins in the remote past. The world beyond the sky has never been very far from the supernatural, and even in our own technological age astronomical objects are regarded with an awe close to reverence. A glance along the shelves in many bookshops will soon reveal the affinity felt by the purchasing public between, for example, black holes and the world of the occult.

The mystical significance of astronomy can be discerned in all ancient cultures. Many monuments, temples, magic charms and decorations, religious relics and documents testify to the powerful influence that events in the heavens have exercised over the affairs of man. In many early cultures, the sky was the domain of the gods, and the organization of the cosmos reflected the metaphysical workings of the supernatural. The astrologers, who could interpret the cosmic order and relate it to the fortunes of mortals, were held in high esteem

1

and enjoyed great political and social power.

It seems unlikely that preoccupation with the cosmic connection emerged merely as a result of institutionalized theology, as it arose among the first great communities at the dawn of recorded history. The common interest in astronomy among such diverse cultures as the Sumerians and the North American Indians points to a far more fundamental awareness of the celestial dimension of mankind. The reason for this remains a mystery, but it must surely be a by-product of our evolution. We may not be the only animals to have noticed the stars, but the long-standing and widespread worship of objects in the sky indicates that astronomy holds a significance for us beyond the simple, practical matter of navigation.

The rise of science and technology as we know them today was closely associated with advances in astronomy made during the European Renaissance of the seventeenth century. Until that time, astronomical pronouncements were largely confined to the priest-hood, as befitted a subject with such direct theological implications. When Galileo, Kepler, Copernicus and later Newton began to analyse the motions of the planets using mathematics and the concept of physical (as opposed to metaphysical) laws, they set into motion a challenge to the religious establishment that was eventually to grow into a completely alternative social philosophy. In the three centuries of turmoil that have followed, science has replaced religion as the dominant force for social structuring. The example of astronomy, to which the application of rational scientific principles produced such stunning success, has been followed by physics, chemistry, geology, engineering and biology.

Few would deny that science works – as an analytic technique for probing inexplicable systems and phenomena, as a framework for understanding and communicating data about the physical world, and as a basis for gaining control over our environment through technology. Yet in spite of the enormous social and intellectual progress that can be directly attributed to scientific advance, many people feel that science and technology lack some of the mystique of the old ways of looking at nature. Varying from vague misgivings to the wholesale rejection of scientific values, modern society, particularly youth, is reacting against science.

Why, in this atmosphere of backlash, does astronomy still exercise the fascination it did over our forefathers? In part this must be due to a vestige of the supernatural that still remains associated with the cosmos. In addition astronomy, of all the sciences, is regarded as the most 'clean'. If you want to blame someone for the atom bomb, industrial pollution, the microprocessor or the extermination of whales, don't blame the astronomer. His discoveries will not threaten your job with new technology or your life with a new weapon. The very remoteness of outer space makes it a safe playground for the inquiring mind.

Whatever the reasons that underlie the tremendous popular interest in astronomy and astrophysics, there is no lack of subject matter to intrigue the devotee. Since the second world war the pace of technological advance and discovery in astronomically based topics has been explosive. First came the development of the radio telescope as a spin-off of radar and radio research during the war. By opening up a completely new window on the universe, radio astronomy enabled objects that had hitherto been unnoticed to be 'seen' by their radio emissions. Radio telescopes pushed back the edges of the observable universe and led directly to the discovery of strange new objects. They also enabled familiar objects to be viewed in a different perspective, and processes to be studied that leave little or no visible imprint.

The Earth's atmosphere effectively blocks all electromagnetic waves except light and radio waves, so that astronomers had to await the development of the artificial satellite before widening the spectrum of wavelengths over which to view the universe. In the 1960s and 1970s whole new branches of astronomy were established: infrared, X ray, and gamma ray telescopes were flown in orbit and sent back a deluge of information about cosmic systems near and far. Further developments in the technology are continuing today, and theorists are hard put to keep pace with the observations. Still more exciting, neutrino telescopes, devices capable of 'looking' into the centre of the sun by detecting one of nature's most elusive subatomic particles, appeared on the scene in the late 1960s. Far from being in orbit above the Earth's atmosphere, the first operating neutrino telescope was installed a mile below ground in South Dakota. Most bizarre of all, astrophysicists began to construct gravity wave telescopes – huge metal bars that are designed to detect gravitational waves emanating from the surfaces of black holes and other exotic objects.

These great advances have shown that the universe is far more complex and bewildering than was suspected a couple of decades ago.

3

They have revealed a cosmos of unimaginable violence and power, yet still one of majesty, serenity and beauty. It is clear that our understanding of the world beyond the sky is in its infancy, and the true nature of man and his relation to the universe is as yet only dimly perceived.

One of the more salutary lessons learned from modern astronomy is that the Earth is utterly insignificant amid the immensity of the cosmos. This is not merely to say that the universe is large – though it is larger than anyone can envisage. The more humbling discovery is that the world we once regarded as the centre of creation is commonplace beyond belief. The Earth, even the sun, seems to be just one of countless billions of other similar bodies scattered without discernible limit throughout space. Our sun is but a typical, ordinary star of which we are surrounded by millions upon millions, grouped together with its neighbours into a gigantic, flattened disc-like structure called the Milky Way galaxy. This galaxy contains about one hundred billion stars, together with a lot of gas and dust, swirled up like a Catherine wheel, and rotating once every 250 million years or so. There is a central spherical nucleus of stars, relatively more crowded, surrounded by several trailing arms wound into a spiral shape. The sun is located near one of these spiral arms.

The Milky Way is big. It takes light no less than 100,000 years to cross it from one side to the other, travelling at the staggering speed of 300,000 kilometres every second. To put this in perspective, light takes a mere 500 seconds to travel the 150 million kilometres from the sun to the Earth. Because astronomical distances are so large, it is convenient to measure them in light years rather than kilometres. One light year is about 9.5 million million kilometres. Our Milky Way galaxy is therefore 100,000 light years in diameter.

Galaxies are the 'atoms' of cosmology – the basic building blocks of the universe. Many billions of galaxies are accessible to modern telescopes; that is, billions of giant structures, each containing an average of a hundred billion stars like the sun. A nearby neighbouring galaxy is the Great Nebula in Andromeda, just visible to the naked eye. This is somewhat more than 1½ million light years away. The most distant galaxies known are several billion light years away, and there is no reason to doubt that if we could probe beyond this distance, there would be more still. Indeed, it may be that there is an infinity of galaxies.

Few people can fail to have been moved by the unfolding panorama that modern astronomy presents, though for all but a tiny number of specialists, who have enjoyed the luxury of many years of challenging study, the new discoveries seem baffling or even incomprehensible. In my own experience lecturing on fundamental astronomy and on cosmology – the study of the overall structure and evolution of the universe – people are passionately interested in black holes, the big bang, pulsars and the like, but have very little idea of what these labels really mean.

In many ways the 1970s were the years of the black hole; the decade even culminated in a Walt Disney film of that title. The public is vaguely aware that black holes are peculiar astronomical objects which suck in material, but do not release it again. The black hole is undoubtedly a fascinating and awesome object, but the hole in itself is really a bit of a scientific red herring. Even before the phrase 'black hole' was coined, astrophysicists were more concerned with the thing inside the hole – the so-called singularity.

What worried the astrophysicists profoundly was the unavoidable prediction, made as long ago as 1939 on the basis of orthodox physics, that a star could collapse without limit until it has shrunk away, quite literally, to nothing. The reason why this issue became of major concern during the 1960s was the discovery that there really do exist stars for which such catastrophic disappearance is likely. As something turning into nothing is without precedent in science, the departure of a complete star from the physical universe is indeed a singular occurrence, so scientists use the word 'singularity' for this demise.

Some physicists regard the singularity as the end of space and time – a route out of the universe into nothing that anybody knows. Others regard it as the disintegration of the known laws of nature. Either way one must concur with the sentiments of John Wheeler, the astrophysicist who invented the name black hole, and who has contributed much to our understanding of both holes and singularities, that the singularity is the greatest crisis that physical science has ever had to face. In the coming chapters we shall see why Wheeler made such a grave pronouncement. The singularity may represent the limits of science itself – the interface between the natural and supernatural.

What, the reader may wonder, makes a star collapse to nothing? The short answer is, gravity. Gravity – the force that keeps our feet on the ground, that controls the motions of the stars and planets, that raises the ocean tides. Gravity is the most familiar and seemingly innocuous

5

of all the forces of nature. Yet it has within its power the ability to smash not only all matter, but space and time as well. It was gravity that, through the proverbial falling apple, dispatched Newton on the road that led to the great scientific revolution of the late seventeenth century.

What could be at once more natural yet so utterly mysterious as a falling apple? Why should the apple fall? Because the Earth pulls it with a force that we call gravity. But how does the apple know that the Earth is there? How does the message 'Fall in this direction' get through to the unsupported apple? Newton made no attempt to explain the mechanism of communication, but merely observed that gravity acts 'at a distance' across empty space without any visible means of contact or influence.

Mysterious or not, the idea is clearly on the right track, for all around us in the universe gravity is reaching out from one body to the next, always remorselessly pulling matter towards other matter. The sun pulls the Earth, the Earth pulls the moon, the moon pulls Neil Armstrong, and so on. All the huge assembly of stars that make up the Milky Way galaxy, of which our own solar system is but a fragment, is bound together by the universal and ubiquitous force of gravity.

We see gravity in action everywhere we look. It is important to realize that gravity not only pulls one body towards another, it also pulls a body inwards on itself. By this ever-present grip, the sun is restrained from boiling away into space, the planets clasp a cloak of atmosphere, the stars cluster into groups, some small, others large. Gravity always has the same tendency: to pull matter together. It is present in every atom, nothing escapes its grip; even the minute force of gravity between lead balls can be measured in the laboratory. The more matter there is, the stronger the pull. Thus the moon, which contains little more than one per cent of the mass of the Earth, has a surface gravity of about one sixth of the Earth's. In contrast, on Jupiter, which has a mass of 318 Earths, an average man would weigh about 200 kg (440 lb).

In fact, the story of the universe is really a story of the struggle against gravity. The Earth holds itself up against its own weight because of the solid material forces (electromagnetic in origin) of the interior. The sun supports its even greater bulk by calling upon immense gas pressure inside its central furnace. Every object that exists has to have a means of avoiding falling together. In relatively

6

low mass bodies, the fight against gravity is easily won. Indeed, gravity is so feeble in our own bodies that we do not notice any force trying to squeeze us together (though we feel the enormously greater force of the Earth's gravity). However, gravity is cumulative, and the greater the quantity of matter the larger the inward pull becomes, until it overwhelms all other forces that nature can muster against it. It becomes all-conquering. When an object gravitates so strongly that it can no longer support its own weight, catastrophe follows, and the body collapses. In the coming chapters we shall follow its fate. For now we merely note that if the sun were ten times as massive (which is not unusual for a star), it could not, in the end, support itself. So gravitational collapse must be taken seriously. We shall see that stars, far from being the natural form of most of the cosmic material, are really only a temporary interlude between the distended clouds of gas from which they are born, and the totally imploded entities – the singularities – in which they die.

Not only stars, but all objects (including our own bodies) are in principle unstable against gravitational collapse. Every body has within it the means of self-destruction. The reason that it is only giant stars that are usually considered in this context is due to a fundamental property of gravity known since the time of Newton. All bodies pull all other bodies with a force of gravity, but the strength of that pull diminishes with the separation between the bodies. The sun pulls on Mercury (the innermost planet) much more strongly than it pulls on Earth. To remain in a stable orbit each planet must experience a balancing force to counteract the sun's inward gravitational pull. Equilibrium is provided by the centrifugal force due to the planet's rotation around the sun. Most of the planets move in almost exactly circular orbits with the sun at the centre. In this symmetrical case, the centrifugal force is equal in magnitude to the sun's gravitational force, but directed the opposite way, i.e. outwards, away from the sun. If the balance between the forces were upset, then the planet would adjust its orbital characteristics until they were restored, e.g. by moving closer to, or farther from, the sun. In the case of Mercury, the stronger pull of the sun must be counteracted by a higher revolution rate to produce a larger centrifugal force. Thus, Mercury's year is only 88 Earth days long. It follows that the measurement of a planet's orbital period (i.e. its year) can be used as a measurement of the sun's gravitational force acting upon it.

A comparison between the orbital periods of all the planets, with their widely different distances from the sun, reveals the precise mathematical way in which the sun's gravity dwindles towards the periphery of the solar system. The result – Newton's so-called inverse square law of gravity – was a pivotal discovery in the progress of science. Physicists and mathematicians have had recourse to this basic law on endless occasions in the three centuries since it was proposed: to study the motions of planets, asteroids, stars and galaxies, to model the evolution of the huge gas clouds found in the galaxy, to model the internal structure of stars, of the sun and of the Earth, to compute the trajectories of rockets and projectiles, and much else.

What does 'inverse square law' mean? It is a way of describing one of the simplest imaginable numerical relationships between the strength of gravity's pull, and the distance from the gravitating object near which the pulled body is located. The relationship says, in essence, that when you double the distance, then the force falls to one quarter of the value it had. Thus, if the Earth were transported to 300 million kilometres from the sun (rather than the actual 150 million) it would feel only 25 per cent of the sun's pull that it experiences now, and would orbit much more slowly (giving us a longer year). Similarly, if the separation of the Earth and sun were increased to three times its present value, the pull would fall to one-ninth; four times and it falls to one-sixteenth, etc. The arithmetic rule should be clear: square the distance and invert the number (hence inverse square law). Thus, five times the separation implies $5 \times 5 = 25$ (squaring five) which leads to a force of $\frac{1}{25}$ (inverting twenty five).

The inverse square law lies at the heart of the extraordinary phenomenon of total gravitational collapse. Although our specific example concerns the sun's gravity acting on the planets, the law is valid for all bodies in the universe – even individual atoms. To appreciate its relevance to collapse, consider what happens as we move, not farther away, but closer in to the sun (or any other body). If the sun's gravity diminishes as a planet's distance from it increases, so the gravity must increase as a planet draws nearer to the sun. By the same, universal, inverse square law the gravity at one-half the Earth's present orbital radius (i.e. at 75 million kilometres) is four times larger ($\frac{1}{2} \times \frac{1}{2} = \frac{1}{4}$; inverting $\frac{1}{4}$ gives 4). At one third of its radius (50 million kilometres from the sun) the gravity is nine times as large, and so on.

In Fig. 1 the force of gravity on a planet is plotted on a graph in terms

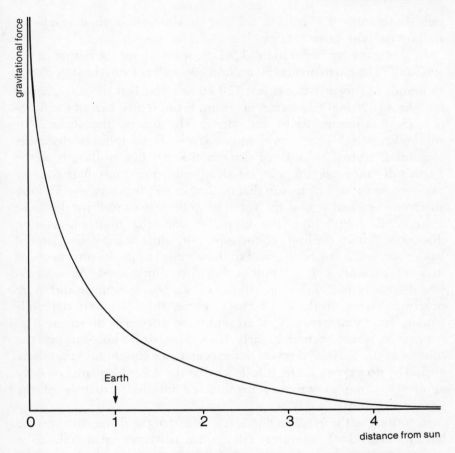

1 Inverse square law. The sun's gravitational force on an Earthlike planet
is plotted against distance from the sun in units of the Earth's orbital
radius. The force rises sharply as the sun is approached.

of the distance separating that planet from the sun, i.e. the radius of its
orbit (assumed circular). A similar curve would apply to the gravitational pull around any other body (e.g. the Earth, a star, a lead ball).
Two features of the curve are of interest. The first is that the gravitational pull falls steadily towards zero at great distance from the sun.
The second is that the curve rises sharply as the sun is approached. In
fact, it goes off the top of our chart. Evidently the gravitational pull felt
as the sun is approached escalates upwards. The inverse square law
shows how: at one hundredth the present distance it rises to 10,000, at

one thousandth it reaches a million, at one millionth it reaches a million million. How far does it go?

At this stage we begin to think a little more about the nature of the sun itself. The Earth orbits 150 million kilometres from the sun, so one millionth of this distance is just 150 kilometres. But the sun is a ball measuring 696,000 kilometres in radius. From which part of the sun is our 150 kilometres to be measured? The centre, the surface, or somewhere else? The inverse square law in fact applies to distances measured from the centre of the sun, so that one millionth of the Earth's distance refers to a region deep inside the sun, where the law ceases to be valid. The reason that the law breaks down when the solar surface is reached is that the force of gravity is caused by the solar material. Each little bit of the sun pulls on all other matter in its own direction. Within the body of the sun itself, some of the solar material lies above, and some below, and so these pull in opposite directions, as in a tug of war. The net pull is therefore diminished. As we pass towards the centre of the sun, the tug of war becomes more and more nearly balanced until, at the exact centre, the solar material pulls equally and symmetrically all around in an outwards direction. The net gravity is zero there. Similarly, if we could drill a hole through the middle of the Earth and travel to the centre, we should find that there would be no gravity there. It follows that the inverse square law only applies so long as we may neglect the internal structure of the gravitating body.

Returning to the graph in Fig. 1, it is clear that the rising curve on the left should level off and start to fall again at a distance equal to the solar radius, because within this distance the above-mentioned effects of the interior solar structure begin. But what if the sun were smaller? Then the curve would rise higher. Let us consider the full implications of this.

The gravitational pull which we have been discussing referred to the force experienced by some external body, such as the Earth. However, the gravity of the sun not only acts on the planets, it also acts on itself. That is, the material of the sun is also pulled down towards the centre of the sun, just as earth is pulled towards the centre of the Earth. If it were not for the sun's gravity grasping at the solar material, the hot gases near the surface would evaporate off into space. According to the inverse square law, the gravity experienced at the solar surface depends on the radius of the sun. At its present radius, the surface

10

gravity of the sun is nearly 28 times that of Earth, so a 100 kg man would weigh 2,800 kg there. But if the sun were shrunk to half its present radius, retaining the same quantity of material (i.e. the same mass), the man would weigh $4 \times 2,800 = 11,200$ kg. The surface gravity increases four times. This means that the weight of its own material depends on how shrunken or distended the sun is.

In the continual struggle between gravity and internal pressure, this dependence of weight on the sun's radius is vital. Suppose the sun were, by some magic, squashed to one half its radius. The self-weight of the sun, i.e. the weight of the solar material in its own gravity, would multiply fourfold. The sun would feel itself to be four times as heavy. This increase in weight would try to shrink the sun still more, leading to a greater increase of surface gravity, hence weight, and still more shrinkage. On the other hand, the compression would also increase the internal thermal pressure, which would fight to support the extra weight by pushing outwards even harder. It is a titanic battle, and one which, in the end, gravity will always win. If only the sun can be shrunken to a small enough radius, no amount of pressure will save it. The critical radius at which gravity is bound to overwhelm all other forces is, however, extremely small – about 1½ kilometres. So long as a star remains hot, it can call upon the thermal pressure to shore it up against the crushing tendency of escalating gravity. But when the star burns out and its reserves of heat and pressure fail, it will succumb to the irresistible force of shrinkage.

The ruthless logic of the inverse square law applies not only to the sun and stars, but to all matter. The Earth, for example, less threatened than the sun because of its lower mass, would nevertheless collapse under its own weight if it were shrunk to the size of a marble. A human being would, if squashed into a ball somewhat smaller than a speck of dust, feel as much self-gravity pulling him together as he feels from the Earth's gravity pulling him down. At a radius of one million-billion-billionth of a centimetre, he would be unable to prevent total implosion under his body's own immense weight, now equivalent to a million billion billion billion billion tonnes.

The simple story of gravity's struggle to compress all matter has been known since the time of Newton. What was not appreciated until this century, however, is the close connection between gravity and the nature of space and time. It is a connection which implies that overwhelming gravitational collapse amounts to much more than a crisis of matter – it is a crisis in the very structure of existence.

There is no doubt that the widespread popular appeal of astronomy is partly the result of the bizarre nature of gravity, in the way it affects space and time. Words like 'space warp' and 'time warp' are commonplace in science fiction and conjure up the impression that gravity can do very peculiar things to those cherished qualities. Although it was Albert Einstein, as long ago as 1915, who originally suggested that space and time could be 'warped' or distorted by gravity, for most people in the 1980s the idea is still little more than a mysterious-sounding phrase. What does a space warp really mean? How can gravity curve space when space contains no substance to be curved?

The basic idea of gravity distorting space and time does not depend on any very advanced or complicated scientific experiments, but can be deduced by the simplest of observations. Indeed, the idea could have been discovered by Newton himself. Newton, and in fact Galileo before him, knew that if two objects were dropped together from the same height they would strike the ground at the same moment. There is a story (probably untrue) that Galileo attempted a demonstration of this type from the famous leaning tower of Pisa to (unsuccessfully) convince his sceptics. Scepticism does frequently greet this claim, because somehow it seems that a heavier body should fall faster than a light one. However, heavy bodies are also more ponderous, which makes them sluggish to respond to the pull of gravity. The two effects, weight and inertia, always compensate, so that the body falls in the same way, whether it be a marble or a cannon ball. Of course, some fluffy objects like feathers fall more slowly, but this has nothing to do with gravity. It is an effect of the air. In a vacuum, even feathers fall as fast as cannon balls. One of the more amusing experiments conducted during the Apollo space programme was for an astronaut to repeat Galileo's experiment in the vacuum conditions of the moon.

The fact that all objects fall in the same way when released means that a person in free fall feels completely weightless. We often get a slight impression of this when an elevator moves downwards suddenly and one's stomach is 'left behind'. Imagine the elevator to be cut loose, and to plunge freely towards the ground. Inside the falling elevator, all objects (people, bags, dogs) fall at the same rate, so they do not move downwards relative to each other. For example, a hapless occupant's pipe, tumbling from his mouth in the surprise of the moment, will fall at the same rate as the man, so will appear to hover, weightlessly, in front of his face. Similar effects are achieved in sky-diving, though in that case air resistance complicates things.

The curiosities of free fall weightlessness never cease to be an attraction. Much of the TV time of orbiting astronauts is used up showing 'floating' antics in space. The orbiting spacecraft is not, as many people mistakenly believe, so far from earth that gravity is negligible. In fact, gravity in orbit is frequently only a few per cent less than at the Earth's surface; some spacecraft are only a few hundred miles above the ground. The reason that the astronauts feel weightless is because they are falling freely; the spacecraft engines are switched off. This sometimes causes confusion because although it is falling, an orbiting spacecraft does not fall to the ground. This is because the Earth is round, and in addition to falling downwards, the spacecraft is also moving sideways at high speed, so by the time it has fallen, say, one kilometre, it has moved so far horizontally that the Earth's curvature has depressed the ground level by a kilometre also. The spacecraft is no nearer the ground at that point than it was before. It just goes on falling round and round the Earth.

There is another way of understanding why a body is weightless in free fall. When an object falls it accelerates downwards. Acceleration, however, produces a sensation exactly like gravity. For example, when a car accelerates forward rapidly, one feels a push in the back. To take another example, on a revolving merry-go-round or a centrifuge, the outward centrifugal force feels like a gravitational force. Often, the force produced by a revolving system is called 'artificial gravity' and is indistinguishable in its local effects from the real thing. There are plans to build a huge space station, shaped like a wheel, slowly rotating to simulate the effect of gravity at the periphery.

Rotation is a form of acceleration, and any sort of acceleration will produce 'artificial gravity'. Therefore, when a body is dropped, as it accelerates downwards it feels a force of artificial gravity acting upwards that exactly cancels the downward pull of the Earth's gravity, producing weightlessness. In the same way, the weightlessness of a body orbiting in a circular path around the Earth can be considered as due to the exact cancellation of the Earth's gravity by 'centrifugal' force.

These considerations of weightlessness in free fall are simple enough, but how do they connect up with space warps? Let us take a closer look at what goes on inside the falling elevator. At first sight

13

everything seems to be, from the point of view of the occupant, floating in its place, maintaining station relative to other things and to the walls of the elevator. But very careful measurements would show a slight, almost imperceptible change. For example, suppose four

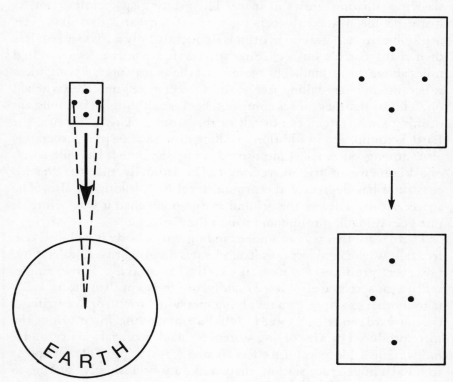

2 Gravitational distortion. As the elevator falls, the four freely falling particles experience slightly different gravity, so the square shape distorts into a diamond as the middle two particles converge on the Earth's centre, while the top and bottom particles drift apart on account of the minute difference in their proximity to Earth. Einstein showed how this geometrical distortion is really due to warped spacetime rather than tiny differences in the gravitational forces.

small objects were arranged in a diamond shape as shown in Fig. 2. As the elevator falls, all the objects fall at equal rates, and the shape does not change. Or does it?

Two objects at the sides of the diamond both fall precisely vertically. But the Earth is not flat, of course, so the direction of the local vertical changes from place to place. Because the objects are separated slightly, the local verticals through each of them will be inclined a tiny amount relative to the others. In fact, instead of falling on parallel paths, the particles all fall directly towards the centre of the Earth. That is, if the elevator were allowed to plunge on down through a tunnel drilled through the middle of the Earth, the two objects at the side of the diamond would converge together there.

Similarly, the object at the base of the diamond is always a little closer to the Earth than its neighbours, so we must take into account the inverse square law which says that there is slightly stronger gravity acting on the bottom object than the others. Consequently it falls slightly faster. On the other hand the object at the top of the diamond trails behind the others. The upshot of all these tiny, but crucially significant, differences is that the diamond shape gradually becomes elongated as it falls until, at the Earth's centre, it flattens completely, as the 'sides' touch.

These small effects are caused by the slight variation of the Earth's gravity from place to place. In free fall, a body is released completely from the sensation of gravity, except for these small variations. For the falling body, gravity only manifests itself at all by the slight changes in position caused by the weightless particles drifting together or apart. Moreover, because *all* bodies, however heavy or light and whatever they are made of, are subject to the same experiences (remember Galileo's experiment), the 'force of gravity', so often discussed without thinking, really begins to seem rather unlike a force at all. In fact, all that gravity really amounts to for a falling body is a gradual distortion of shapes. Drop a diamond and it becomes elongated, drop a flexible ring and it becomes squashed into an oval.

All of these observations suggest that we should stop thinking about 'the force of gravity' altogether and use the language of *geometry* instead to discuss gravitational effects. This was Einstein's great conception. Gravity can be abolished as a force and replaced by geometry. Gravity *is* geometry. This is a bold, almost outrageous suggestion, but the detailed theory of gravity as geometry has been tested again and again. No alternative theory that does not use this central idea has survived the rigorous scrutiny of modern experiments.

What, then, is a space warp? The falling diamond is warped or distorted in shape. This phenomenon is hardly a property of the diamond itself, for a similar distortion would be achieved by any other falling object. It must instead be regarded as a property of space. It is space that is distorted; the diamond only glides freely through a curved space.

We can begin to glimpse the idea of a space warp, though it still seems preposterous. How can space bend? Space is emptiness; emptiness cannot have shape. A close analogy can help throw more light on what is really meant by curved space. Consider two aircraft ten kilometres apart at the equator. The pilots are given instructions to fly exactly due north. The planes move off on parallel paths and the passengers settle down for a long flight. After a while they notice that the aircraft seem to be slightly closer than they were before. This is baffling because they started out flying exactly parallel, and each pilot is flying precisely due north, without deviation. Nevertheless, as time goes on the planes get closer and closer together until, at the north pole, disaster strikes – the planes collide. Pilot error is ruled out, for both pilots kept on flying precisely due north. What has happened?

The answer is to be found, of course, not in any mysterious force that draws the two aircraft together, but in the curvature of the Earth's

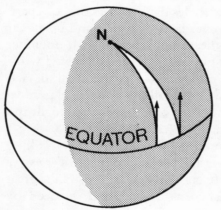

3 Two lines of longitude, exactly parallel at the equator, nevertheless converge and touch at the north pole. Two aircraft, flying precisely north along these lines, appear to be 'forced' gradually together. The explanation, however, is not the presence of forces, but geometrical distortion due to the underlying sphericity of the Earth.

surface (see Fig. 3). Although the lines of longitude are exactly parallel at the equator, they do not remain so, but slowly converge towards the north pole. At the north pole these initially 'parallel' lines cross each other. By flying up the parallels, the pilots were assured of a polar disaster. On a curved surface, parallel lines can intersect.

The situation is reminiscent of our falling diamond. Rather than invent a mysterious force (gravity) to explain why the objects at the sides of the diamond slowly converge, we should rather say that the objects are 'flying', force-free, through a curved space. The centre of the Earth is like the north pole for the pilots. It is where the edges of the diamond collide. And the distortion or warping of space is rather like the distortion or warping of a map of the world in Mercator's projection, which can only depict the spherical Earth on a flat sheet by stretching distances greatly near the polar regions.

Actually, the warp in the case of gravity is not, strictly, a space warp, but a spacetime warp. Gravity affects time as well as space, and it turns out that space and time are really in any case closely interwoven, as we shall see in chapter 3.

A good place to look for a space warp in action is around the sun, which is the biggest gravitating object in our vicinity of the universe. An attempt at such was made as long ago as 1919 by Sir Arthur Eddington as a crucial check on Einstein's theory. The idea is that as the sun moves across the sky (due in reality to the Earth's motion) it slowly wanders through the constellations of the zodiac. When seen close to the sun, the solar space warp distorts the shapes of the constellations slightly by displacing the apparent positions of the stars somewhat, just like a lens. This effect shows up during a solar eclipse, when the moon blots out the sun's glare and enables the stars to be seen in day-time. A careful comparison of the positions of stars that appear close to the sun (i.e. near our line of sight to the sun) with their recorded positions when the sun is in another part of the sky, does indeed reveal the warp effect of the sun's gravity (see chapter 3).

How is one to visualize such a peculiar and unfamiliar effect as warped space? The Earth's surface gives a clue. We can happily visualize a bent, curved or warped surface (see Fig. 3); think, for example, of the surface of a balloon, or the topography of a mountainous country. However, space is not a surface. It is three-dimensional, for a start. This makes it hard to imagine what distortion and curvature looks like, but the basic idea obviously does not depend on the dimensionality. We can have curvature in one dimension (a curved line) and two dimensions, so why not three, or even four (spacetime)? All it means is that the usual rules of school geometry (e.g. angles of a triangle sum to two right angles) are no longer exactly true, but simply approximations. To aid visualization, we shall often depict space as a

17

surface rather than a volume. This is a simplifying device no more remarkable than drawing a two-dimensional section through a house, or a piece of machinery, as is done so often by architects and engineers.

There is still a feeling of unease about curved space. Granted that three-dimensional curvature makes sense, what exactly is it that is curved? The surface of a balloon, for example, is a membrane of rubber. Space is nothing at all.

Perhaps one of the great scientific discoveries of the twentieth century is that space is indeed 'something'. In many ways it resembles a block of rubber, filling the chasms between bodies. (We shall therefore often think of it as a membrane of rubber in our two-dimensional view.) Although we cannot feel or touch space, in the sense of a substance, it can nevertheless be bent, stretched and twisted. We shall see in later chapters how, for example, intergalactic space is steadily stretching. We shall find that during catastrophic gravitational collapse, space can bend so far that the 'rubber' actually snaps and space comes apart at the seams. We shall also find that time can be stretched in this way, and may also 'snap' and break off.

Once it is accepted that space can be curved, an exciting new possibility arises. Let us, as promised, think of space now as a two-dimensional sheet or membrane, and envisage the experiences of a two-dimensional 'pancake' creature who crawls about on the surface. If the pancake is small he will not notice the lumps and bumps locally, because in a small enough region the space appears flat (just as the curvature of the Earth is only apparent on a large scale). He will probably, if he hasn't read Einstein, think that his universe is an infinitely extended flat sheet of space. On closer inspection, the pancake-creature finds it to be a bit bumpy, so he envisages a bumpy sheet, stretching out for ever in all directions. Then he makes a shocking discovery. He is living, not on an infinite sheet, but on the surface of a bumpy balloon. His universe, far from stretching to infinity, curves round and joins up with itself. Not only is his space bent, but it is self-connected and finite in area. The pancake-creature visits all places – he 'does' the complete pancake cosmos. Nowhere does he encounter an edge or boundary or barrier where the space 'stops'. Yet it is still only finite in extent. He can travel right around it and return to his starting place.

Is our universe like that – curved right around so it connects up with itself? Is it finite in volume, yet with no edge or barrier? Perhaps.

These issues will be examined in detail in the later chapters.

Mathematicians have invented a whole subject solely devoted to the analysis of the way that lines, surfaces and volumes connect up with themselves and each other. It is called topology. A topologist is not interested in detailed geometrical information such as how big something is, or how many bumps it contains of a certain size, but only in such issues as whether a two-dimensional surface is infinite, or closed like a balloon, or whether a string is knotted or unknotted.

We shall often have occasion to use topology in this book, so some general ideas will be presented here to give something of the flavour of the subject. Consider first two familiar objects, a doughnut and a

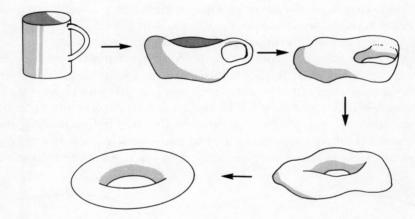

4 The topology of the cup and doughnut are the same, because their shapes may be continuously deformed into one another without tearing or rejoining.

teacup (see Fig. 4). The shapes of these two items are very different, but they do have one vital feature in common: both have a 'hole' in them. In the case of the doughnut the hole goes right through the middle, while for the teacup the hole is in the handle. Nevertheless, they share a property that is not possessed, for example, by a potato (no hole) or a two-handled jug (two holes). We say that the doughnut and the teacup have the same topology, which in turn differs from the topology of the potato or the two-handled jug. Notice that the shapes or sizes do not affect this designation. In fact, if we envisage these items as all made out of putty, then the teacup can be continuously deformed into a doughnut purely by stretching, bending and squeezing. There is no need to tear or join together separate regions of the putty. This would not apply if we wanted to end up with the potato shape, or that of a two-handled jug.

The topology of surfaces is also an intriguing subject. A famous long-standing problem, only recently solved, is the so-called four colour problem. This has to do with the minimum number of colours necessary to colour countries on a map so that no two adjacent countries need have the same colour, however complicated the map may be. Once again, the precise shape of the countries is immaterial; what counts is their overall relationship to one another. Cartographers have long known that only four colours are necessary in practice, but proving that there can never be a map complicated enough to require five or more colours has taken many years of effort.

Another amusing product of the topological investigation of surfaces is the so-called Möbius strip. A strip of paper can have its topology changed by folding it into a loop and joining. The loop will have the geometry of a cylinder, with two edges (top and bottom) and two sides (inside and outside). If, instead of looping and joining in the ordinary way, a single twist is put into the loop before joining, one obtains a 'kinked' loop – the Möbius strip – which has only one edge and one side, as may easily be verified by tracing a route around the

 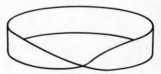

5 The topology of the loop on the left is drastically different from the Möbius strip on the right. The latter has only one side and one edge.

strip (see Fig. 5). The cylindrical and Möbius strips have quite different topologies. Analogous 'kinked' or 'twisted' structures can be constructed in higher dimensions.

A final example of a problem in topology concerns a conundrum presented to the brilliant Swiss eighteenth-century mathematician Leonhard Euler. The city of Königsberg was in those days built around an island situated at a point of bifurcation of the river Pregel. The various districts of the city were connected by seven bridges as

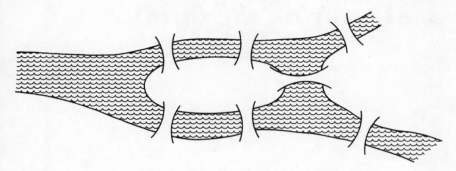

6 The seven bridges of Königsberg. Is it possible to cross each bridge in turn in one unbroken path without recrossing? This classic problem of topology was solved by Euler in the eighteenth century – it is impossible.

shown in Fig. 6. Afternoon strollers were intrigued by the conjecture that it was possible to cross all the bridges on a continuous walk without crossing any one of them twice. Here again we have a problem of topology, for the sizes or shapes of the bridges are irrelevant. Only the way in which they connect with each other is at issue. Euler proved the conjecture was false.

It may seem that the Möbius strip or the bridges of Königsberg are a world away from the cosmic connection, but that is not so. Outer space is the testing ground for Einstein's theory of curved space and time. It is among the unimaginably powerful gravitational fields of distant astronomical bodies that space is buckled and bent, perhaps even torn and joined in Möbius forms, or more complicated topologies. We shall see, in the coming chapters, that topology plays a central role in the prediction of nature's most bizarre structure – the naked singularity.

We have surveyed modern ideas of astronomy and gravity, and found that Newton's inverse square law leads to the possibility that gravity can overwhelm all other forces, causing a star to implode on itself without any hope of support. When this happens, the graph in Fig. 1 goes on climbing higher and higher without limit. To where does it climb? To infinity. The infinite is what follows the triumph of gravity. It is therefore to the study of the infinite that we now turn.

2 Measuring the infinite

To many people the infinite implies something immense and unknowable. In popular language the word is often used loosely to denote 'exceedingly large' or 'beyond anyone's capacity to count'. Often cited is the number of stars in the sky or grains of sand on the seashore. These examples are not, of course, truly infinite; only about two or three thousand stars are visible to the naked eye at any one time. In fact, in daily life we never have occasion to encounter the infinite.

In science, however, infinity is frequently encountered, sometimes with dismay. Long ago mathematicians began attempts to get the measure of the infinite and to discover rules which would enable infinity to join the ranks of other mathematical objects as a well understood and disciplined logical concept. They were in for many surprises. The classical Greeks made only limited progress, and it was not until the nineteenth century and the work of powerful mathematicians like Georg Cantor and Karl Weierstrass that decisive progress was made.

Even in science, for many purposes, infinity is only an idealization for a quantity which is actually so large that to treat it as strictly infinite involves negligible error. From time to time, though, the appearance of infinity in a physical theory denotes something much more dramatic – the end of either the theory, or the subject of its description. This is the case with spacetime singularities. There we are brought face to face with infinity, and it seems to be telling us something profound: that we have reached the end of the universe.

To understand the full implications of singularities, it is first necessary to feel at home with the infinite. In this chapter some of the basic concepts will be explained in easy language. None of the results quoted will be rigorously proved, for the proofs would require many years study of advanced mathematics to comprehend. It is important to realize that the subject of discussion is not a theory about the world, but mathematics. Given the fundamental axioms on which all mathematics ultimately rests, the results are therefore correct, beyond any possibility of doubt, as all the proofs rest on concrete and universally accepted logic. This point is stressed because the results often seem impossible to believe; yet they are true. We shall see that measuring infinity can be a very strange experience indeed.

The first step on the road to infinity is to discard any ideas about 'very, very large'. Infinity is larger than any number, however large that number may be – and there is no limit to numbers. We shall see that not only is infinity beyond all limits, but is, in a sense, so large that it is almost impossible to make it larger. An infinite collection of things obviously cannot be written out, so it will often be denoted by three dots as, for example, the infinity of natural numbers 1, 2, 3, ... The dots mean that there is no right-hand end to the sequence.

The concept of infinity has been a source of confusion and anxiety for over two millennia. The Ancient Greeks made heroic attempts to get to grips with it, but often uncovered paradoxes or arrived at absurd conclusions when misled by common-sense intuition. The subject frequently became embroiled in philosophical and theological controversy, and in 1600 even contributed to the death sentence passed on Giordano Bruno at the hands of the Church. Bruno had declared a belief in the infinity of worlds, against the established doctrine that only God was infinite.

Many people first encounter the idea of infinity when thinking about the universe. Does it extend for ever? If space is not unlimited in extent, does that not mean that there exists a barrier somewhere – in which case the barrier must lie beyond, and something beyond that . . . ? Another question, frequently asked by children, is of the 'what happened before that' variety. It seems that every event must have been preceded by some cause, and every elapsed moment must have come after an earlier moment. We shall see that the answers to these questions can be bewilderingly different from the obvious.

The Ancient Greeks encountered infinity not only in contemplating the limits or otherwise to space and time. In their mathematics too, it was often necessary to consider quantities growing without limit. The

23

circle, for example, was considered to be the product of a polygon for which the number of sides grows infinite. The idea is depicted in Fig. 7. The pentagon is a better approximation to the circle than the

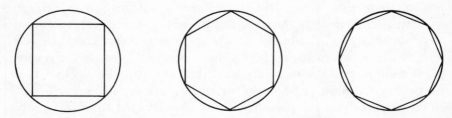

7 As the number of sides is increased, the inscribed polygon approximates more closely in shape to a circle. In the limit of an infinitely-many sided figure, the polygon and circle coincide.

square, while the octagon is better still, and so on.

Considerations such as these led to the concept of infinity as something to be approached – the end goal of an unending sequence of operations or successive approximations. Aristotle spoke of 'potential infinity', an idea that has recurred throughout the centuries, as something for ever approached but never accomplished. In the nineteenth century Emmanuel Kant echoed Aristotle in pronouncing that 'an absolute limit [i.e. reaching infinity] is impossible in experience', while no less a mathematician than the great Karl Friedrich Gauss protested in 1831 about the 'use of the infinite as something consummated'. Even in this century, Hermann Weyl, a mathematician and physicist of immense distinction, would not countenance the notion of infinity as a completed thing. 'It remains', he says, 'for ever in the status of creation', but never achieved.

The basic principle behind potential infinity is that there are systems, such as the natural numbers, 1, 2, 3, ... , that have no upper bound, so that nothing prevents them from growing ever larger. Yet infinity itself is clearly not a number, or anything like it. Infinity, for the adherents of these views, is something one always approaches but never reaches – a sort of open-ended arrangement.

That infinity should stand like a forever unattainable goal, never 'consummated', leads to some uncomfortable paradoxes, as first demonstrated by Zeno of Elea in the fifth century BC. Consider, for example, the curious experiences of an arrow that travels towards a target 200 metres away at 100 metres per second. Common sense tells us that it reaches the target after two seconds' flight. But before the target can be reached the arrow must reach the middle of its flight; this takes 1 second. This much accomplished, it must then reach the midpoint of the remaining gap, which takes ½ second. Travel to the midpoint of that remainder takes a further ¼ second, and so on. The number of additional increments (albeit of steadily diminishing magnitude) is without limit. How, then, can the arrow ever reach its target if infinity is something that can never be accomplished?

The uneasy feeling which this paradox creates plagued mathematics for 2,000 years, and shows how slippery the concept of infinity can be. However closely the arrow approaches its target, there are still more steps ahead of it than behind it. The example also illustrates what is sometimes considered surprising: that the addition of an infinity of numbers can nevertheless produce a finite answer. In this case, the two-second flight time is composed of the union of all the sub-increments:

$$1 + \tfrac{1}{2} + \tfrac{1}{4} + \tfrac{1}{8} + \ldots = 2.$$

Many paradoxes of the infinite were resolved, at least to the satisfaction of some mathematicians and philosophers, only with the work of Cantor and Weierstrass in the nineteenth century. The way was paved by Bernhard Bolzano, a Czech theologian and amateur mathematician, who attempted to lay firm, logical foundations for the concept of an 'actual infinite'; that is, infinity as a completed, realized entity, and not merely a goal, that is approachable but for ever unattainable. For his pains he was forbidden in 1819 to teach or publish in his capacity as Professor of Religious Philosophy at the University of Prague. His main treatise, *Paradoxes of the Infinite*, was published posthumously in 1850. By elevating infinity from a potential goal to an accomplished thing, mathematicians would be able to manipulate infinity with confidence, according to well-understood rules, just like any other mathematical object.

The decisive steps forward were taken by Georg Cantor, one of the strangest figures in the history of mathematics. Born of Danish parents in St Petersburg, Russia, in 1845, he lived and studied mainly in Germany, never making much impact on the academic world. He spent all his professional career at the little-known University of Halle. Yet his achievements were far-reaching. Credited with the invention of so-called set theory (now routinely taught in junior school), Cantor established the theory of 'transfinite cardinals' which has become the basis of all subsequent work on the notion of infinity.

The essence of Cantor's approach to infinity had been around in vague form for centuries. For example, the following curious fact was noticed as long ago as the seventeenth century by Gottfried Leibniz. Imagine all the natural numbers standing in a row: 1, 2, 3, ... The order is given 'Double your value!' and they all turn into even numbers 2, 4, 6, ... Clearly this mutation has not changed the number of numbers standing there, so there must be as many even numbers as there were natural number originally. But the original set of numbers 1, 2, 3, 4, 5, ... *contains* the even numbers, and the odd ones too! Evidently the set of natural numbers is so big that its members are as numerous as only part of itself (see Fig. 8).

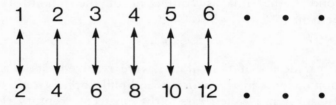

8 Infinity is as big as only part of itself. Every whole number in the (unending) top row can be paired off with an even number in the unending bottom row. None is left over without a partner. Yet all the numbers in the bottom row are contained in the top row, with more besides!

The conclusion seems so stunning because one intuitively expects there to be only half as many even numbers as even plus odd. But when the sets are infinitely extended intuition misleads us. The correctness of the result is most convincingly demonstrated by envisaging all the whole numbers written one at a time on an unending line of blank cards. On the reverse side of each card is written that number which is twice the value on the front side (e.g., card 6 in the row has 12 on the back). The massed rank of cards now displays the infinity of whole numbers. But an observer on the far side of the cards would see instead the unending array of even numbers 2, 4, 6, ... As both observers obviously see the same cards, the even numbers on the reverse side must be as numerous as both the even and odd numbers on the front. Yet the same array of even numbers could have been seen from the front by knocking down all the odd cards, so we are forced to conclude that knocking out every other card does not make the total number of cards less numerous than before. Conversely, the infinite set of all even numbers is made no bigger (i.e. no more numerous) by adding all the odd numbers to it.

This is surely an astonishing result. Is there a logical flaw? Apparently not, for if it were true that there were *more* natural numbers than even numbers, then we could not pair them all off one by one with the evens. There would be whole numbers (e.g. odd numbers) left over without any corresponding even numbers to match them. But every number on the front of a card, however large, has a corresponding even number on the back to pair with, so every whole number, odd as well as even, can be accommodated in a one–one pairing scheme with the even numbers alone.

If the infinity of all even numbers is as numerous as all the even and odd numbers together, it looks, crudely speaking, as though doubling infinity still leaves us with the same infinity. Moreover, it is easily shown that trebling, quadrupling or any higher multiplication of infinity has equally little effect. In fact, even if we multiply infinity by infinity itself it still stubbornly refuses to grow any larger. The square of infinity is only as numerous as the natural numbers.

This result can be demonstrated by envisaging an infinite set of beads arranged in a row, and then another infinite set arranged in a second row above, then a third and a fourth and so on, without limit. We arrive at a lattice or matrix of beads (see Fig. 9(i)) containing an infinity of rows and an infinity of columns – an infinite square (which is, of course, the square of infinity). Now one can easily show that there are no more beads in that unending square lattice, with its infinitely long columns, than there are in the single (unending) row of beads at the bottom. Preposterous?

To see that this is in fact the case we need only glance at Fig. 9(ii). Follow the zig-zag path along the arrows. Clearly it will eventually reach every bead in the lattice. Now as each bead is passed en route, tick off one of the beads in the bottom row. One by one the lattice beads pair off with the beads on the bottom row. No beads in the lattice will remain unpaired, so there *must* be as many beads in the whole lattice as there are in the bottom row of the lattice. What an extraordinarily counter-intuitive result this is. The single row of

27

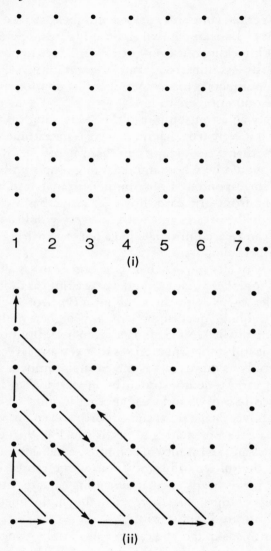

(i)

(ii)

9 Infinity × infinity = infinity. This baffling equation can be seen to be correct by following the zig-zag path through the unending lattice of beads. Each bead may be counted (1, 2, 3 . . .) and paired one by one with, say, the beads in the bottom row alone. For every bead in the lattice above, there is a partner to be found in the bottom row. Conclusion: there are no more beads in the entire infinitely tall lattice than there are in just one of its rows.

bottom beads looks every bit like a tiny (indeed infinitesimal) fraction of the total vast array, stretching as it does to infinity vertically. Yet concrete logic informs us that the infinite lattice is no bigger than just one of the infinite number of its component parts. Moreover, the result can be extended to a cubic lattice and even beyond. In fact, however many times we multiply infinity by itself, we do not succeed in making it any bigger whatever.

If it seems surprising that amassing all the odd numbers with all the even numbers does not make infinity any bigger, it will seem more remarkable to find out that the result remains true even if we throw in all fractional numbers as well. At school we learn to write fractions as quotients of whole numbers, e.g. 3/5, 91/217, 10514/69393. As the whole numbers are limitless and inexhaustible, so are the fractions that can be formed from them. But more than this, for there are whole cohorts of infinities of fractions. This is because between any two fractions, however close in value they may be, there are still an infinity of other fractions. For example, take the two nearby fractions 1/250 and 1/251. Between them lie other fractions, such as 2/501 and 4/1001. Between these two lie others, and so on, without limit.

One can also understand this property visually, by setting up a one-one correspondence between numbers and points on a line, as we do when using a tape measure marked out with numbers. In Fig. 10 the

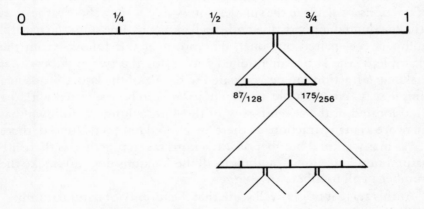

10 A continuous line may be subdivided without limit. Each segment can be blown up to reveal an endless sequence of sub-segments within it, each bursting with infinitely densely packed fractions. Yet there are no more fractions than natural numbers 1, 2, 3 . . . .

points 0, 1/4, 1/2, 3/4 and 1 have been marked on a line, representing the distance of the point from the left-hand end of the line. A 'point' here is supposed to mean a location on the line that has no actual size or extension whatever; something with zero area – think of a little dot shrunk down until its boundary vanishes. Naturally, such a vanishingly small entity cannot be depicted in Fig. 10, so a crude mark is used to denote its approximate position.

If one envisages two very nearby fractions, then they will correspond to two very closely situated points (dots) on the line. The points will sandwich between them a strip or segment of line which has a very small length. Taking a magnifying glass, one can blow up this segment and see that it can be divided into sub-segments, for within its length there are an infinity of points and an infinity of fractions. We can go on magnifying the segment, seeing all the sub-sub-divisions, without any limit. Every time two distinct fractions are chosen, however close their corresponding points may be located, we can always blow up the scale and find endless other points (fractions) between them. And this is true along the whole of an infinitely long line, stretching all the way off to the right through all the whole numbers 1, 2, 3, ... There are thus infinity upon infinity upon infinity of fractions, bursting limitlessly out of ever-smaller segments of the line. But ask a mathematician whether there are more fractions than whole numbers and he will say no. In spite of the infinite reserves of fractions available in the smallest intervals one may choose, they can still be counted individually and labelled one by one with the natural numbers (i.e. paired one-one). The proof of this follows from the example of the lattice mentioned above, for the two numbers that make up a fraction can be associated with one of the lattice beads in a simple way. For example, the fraction 7/43 can be associated with the bead located in the seventh row of the 43rd column. It follows that there are as many fractions as there are natural numbers. Indeed, there are as many natural numbers *and* fractions taken together as there are natural numbers alone – adding in all the fractions does not make the number of all numbers any bigger.

At this stage it might well seem that the infinity of natural numbers really is so big that it cannot be made any bigger, but this is wrong. In a celebrated theorem, Georg Cantor proved the seemingly impossible – that there are infinite sets which are so big that their elements cannot be counted, even with the infinity of natural numbers at one's disposal.

This must therefore be an infinity bigger than all the natural numbers and all the fractional numbers (which can, as we have just seen, be counted) taken together.

Now this result may be hard to swallow because many people believe that the natural and fractional numbers must be all the numbers that there are. How can there be more of anything than there are numbers? However, it has been known since the time of Pythagoras that there exist numbers that cannot be expressed as either a whole number or as a fraction – an infinity of them, in fact. Before discussing Cantor's proof, let us take a look at this mystery of the missing numbers.

Glance again at Fig. 10. Every fraction or whole number corresponds to a point on the line. But what about the converse? Does every point have a fraction to label it? The Ancient Greeks once thought so. After all, there are endless armies of fractions available, packed infinitely densely. Infinitely, though, is a capricious concept, so we have to take care. Are there points on the line that are somehow 'hiding between' the fractions, even though the fractions are infinitesimally close to their neighbours? If there exist, in some strange sense, 'gaps' between the fractions, we know that these gaps must have zero size. Yet a point also has zero size, so it is certainly conceivable that there are points that have been overlooked by the fractions.

A colleague of Pythagoras came up with a simple proof that there are indeed extra numbers 'in the gaps'. Take a square, he said, with side measuring one unit (one metre, if you like). Then the length of the square's diagonal (known from practical measurement to be *about* $1^{207}/_{500}$) cannot be expressed as an exact fraction. Put differently, the point on our line in Fig. 10 corresponding to the length of the square's diagonal has no fraction labelling it. It lies, somehow and incredibly, in an infinitesimally small 'gap' between the infinitely densely packed fractions near $1^{207}/_{500}$.

The proof of this remarkable assertion is always a source of fascination to school children. The length of the diagonal concerned can be expressed as the square root of 2 (written $\sqrt{2}$) because it was known from Pythagoras' theorem that when multiplied by itself, this mysterious new number gave 2 (i.e. $\sqrt{2} \times \sqrt{2} = 2$). The proof consists of showing that $\sqrt{2}$ cannot be expressed in the usual fractional form n/m, where n and m can be any whole numbers without a common divisor (except 1). It proceeds by contradiction; that is, by assuming that $\sqrt{2}=n/m$ and then showing that there are no numbers n and m that will do the trick.

If $\sqrt{2}=n/m$, then squaring both sides of this equation yields $2=n^2/m^2$, i.e., $n^2=2m^2$. Clearly, therefore, n^2 is an even number, as it is twice the whole number m^2. But if n^2 is even, so must n be even (an odd number when squared gives another odd number). Here is where we begin to see something peculiar going on, because all even numbers are twice another whole number, so we could write n as 2p, say, where p is some other whole number. Doing this, we find $n^2=4p^2$, so our equation $n^2=2m^2$ now becomes $4p^2=2m^2$, or $m^2=2p^2$. But we can now argue again that m^2 (hence m) must be an even number, being twice the whole number p^2. The conclusion to this piece of straight-forward reasoning is that both n and m are even, in obvious contradiction to the fact that they have no common divisor: they can both be divided by 2.

The Greeks found other examples of 'gap' numbers. One of these, denoted π, is the ratio of the circumference to the diameter of a circle. Early suspicion of these fractionless numbers led to the appellation 'irrational', the remaining, honest, fractions and whole numbers now being described as 'rational'. In order to symbolize the full panoply of rational and irrational numbers the decimal system must be used. While every fraction can be expressed as a decimal, the converse is not true. Some fractions, like ¼, have finite decimal forms (0.25), though others, such as ⅓, need an infinite decimal (0.3333 ...). All irrational numbers, such as π, need infinite decimals (3.142 ...).

Cantor's great discovery was that the set of all decimals (i.e. all rational and irrational numbers) is a bigger infinity than the set of all fractions (i.e. rational numbers alone). These issues may appear to be mathematical quibbles, but they run very deep. Centuries of groping towards a proper understanding of time, space, order, number and topology lie behind the work of Cantor and others to grasp the infinite as an actual, concrete concept. Some of the greatest minds in human history have foundered on the rock of the infinite. Few ideas can have so challenged man's intellect.

The essence of Cantor's proof is that, if the decimals were to be only as numerous as the fractions, which in turn are as numerous as the natural numbers (as we have seen), then it must be possible to count or label all the decimals one by one with the whole numbers, 1, 2, 3 ... .

0.**2**8307149...
0.9**1**521932...
0.88**4**75628...
0.310**7**8454...
0.2913**9**266...
0.76842**0**31...
0.419866**5**3...
0.600279**3**8...

11 The decimals between 0 and 1 cannot be counted, even with all the infinity of natural numbers, 1, 2, 3 . . . . If each bold-faced digit is altered by 1 then the resulting diagonal decimal cannot be present anywhere in the original list, however many decimals are present in our list. Conclusion: there are more decimal numbers between 0 and 1 than all the fractions.

This means that if all the decimals are written out, one under the other in an infinite column, we could label them 1, 2, 3, ... (see Fig. 11). The particular ordering does not matter, and Fig. 11 just shows some random choice. Cantor pointed out how one can now construct another decimal that cannot be present anywhere in this column, thus contradicting the assumption that all the decimals can be enumerated one by one.

To achieve Cantor's construction, simply go down the diagonal and pull out each digit entered there, rewriting them in a row thus: 2 1 4 7 9 0 5 8 ... . Then change this array by subtracting one from each entry (adding one in the case of zero): 1 0 3 6 8 1 4 7 ... . Clearly, the decimal 0.10368147 ... cannot be present *anywhere* in our original column,

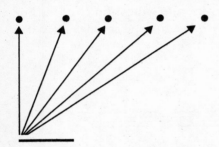

12 Two orders of infinity. Imagine attempting to dismember a continuous line by taking points, one by one, from the line, starting at the left end, and arranging them individually in a row as shown. Even when one has removed an infinity of points, the line has not shrunk by even the smallest distance, for the isolated points, each of zero length, can never accumulate to measure any length at all. Similarly, an infinity of isolated points, if shrunk down together without limit, can never fill out a continuous line, however short one is prepared to accept.

because it differs from every entry there in at least one of the decimal places, due to the above adulteration. The conclusion is stunning. There are an infinity of rational numbers and an infinity of decimals, but the latter infinity is *bigger* than the former.

Armed with this amazing fact, we can now see an important property about the representation of numbers by points on a line. Consider the row of dots in Fig. 12. This row extends infinitely far to the right. Imagine that the row could be shrunk to the left, faster and faster, until all the gaps between the dots vanish away to nothing. We now have what would seem to be a 'solid' line. But this cannot be so. The dots can be labelled by the integers, so their infinity is of the 'smaller' variety. Even though we pack them together infinitely densely, there are not enough to fill out a *continuous* line. For that we need the 'larger' infinity – the infinity of all decimals.

One way of describing the difference between the infinity of dots squashed together and the continuous line is in terms of length. Each dot, by definition, has zero length. So any number of them, even an infinite number, still has zero length. We need many more dots (or points) than an unending row of them to fill out the continuous line. The line clearly does have length.

These two orders of infinity are sometimes distinguished by calling the former discrete, or countable, meaning that, crudely speaking, the elements or points are disconnected from each other and can be distinguished one at a time. The other infinity – the infinity of the line – is called continuous. This distinction becomes very relevant when one considers the nature of space and time.

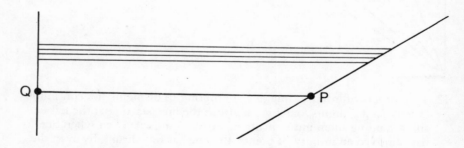

13 The two lines, though of different length, possess the same number of points, for each point P on the oblique line can be uniquely paired with a point Q on the vertical line.

It should now come as no surprise to learn that you cannot make a continuous infinity bigger by any straightforward means: for example, by extending the length of the line. A line one centimetre long has the same number of points as a line two centimetres long. This is easily seen to be correct by glancing at Fig. 13. The oblique line is twice as long as the vertical line, but every point of the oblique line can be joined to one and only one point of the vertical line by drawing parallel construction lines horizontally between them. At first it seems that there must be points 'left over' on the oblique line, but clearly this cannot be the case. Because the horizontal parallel lines can never intersect each other, every point on the oblique line must have its very own partner on the vertical line.

It is but a small step from the observation that there are as many points in a short line as there are in a long one, to the fact that there are no more points in an *infinitely* long line. This construction is shown in

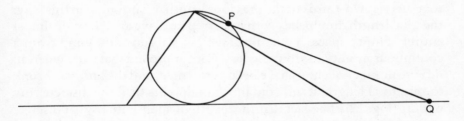

14 The circle contains as many points as the line (imagined to stretch either way to infinity). Each point such as P on the circle can be paired one–one with a point Q on the line. (The single point at the top of the circle requires special consideration.)

Fig. 14, where every point in the circle is joined to every point on the straight line, which must be imagined as stretching on for ever. More than this: just as we can 'square' discrete infinity and still get no more points (Fig. 9) so we can 'square' a line to obtain a sheet or plane, infinitely extended, and not make our continuous infinity any larger. We can even 'cube' it and consider the entire volume of infinite space. This unending universe has no more points than the short line shown in Fig. 12!

If this result seems bewildering, worse is to follow. We might wonder just how short our 'short line' can be, and still contain as many points as all of infinite space. As a short line is as good as a long line in this respect, we can make it as short as we please. In fact, we can make it *no length at all* and still have enough points to match, one for one, all points of infinite space.

This mind-boggling assertion was also proved by Cantor. His construction is shown in Fig. 15. Take a line of unit length. Then chop

0   ⅓   ⅔   1

15 Cantor's set. By successively deleting the middle one-thirds of each line segment, one approaches a collection of line fragments whose total length is zero, yet whose points are as numerous as the entire universe. The third stage of the sequence is shown.

out the middle one-third. Then chop out the middle one-third of the two remaining pieces, then the middle thirds of those remainders, and so on. With each step, the total length of line (adding all the bits together) is two-thirds of the previous length. Continuing in this way, the total length diminishes with each step until we are left with almost entirely empty space. Yet at no stage can we suddenly jump from a continuous infinity to anything less. Cantor proved that even when an infinite number of chops have been performed, and the line has shrunk to zero total length, it *still* contains more points than the discrete row of dots (Fig. 12). The fact that the length of an infinite row of discrete points is zero, and the length of Cantor's set of points is zero, does not mean they contain the same *number* of points. The extraordinary conclusion is that a line of zero length can contain as many points as the whole universe.

It is helpful to understand the relation between infinity and its, also somewhat mysterious, opposite: zero. In a sense, zero is infinitely small, the opposite limit to infinity. Problems occur if zero and infinity are multiplied, and such operations must be handled with care.

Infinity can be generated by zero, and vice versa, in a very simple way. When we write one-half as ½, we mean 'one divided by 2'. Similarly one quarter is one divided by four, ¼, and so on. The larger the number on the bottom, the smaller the fraction obtained. However small a number y we choose (i.e. however close to zero we get) we can always find another number, x, which is large enough such that $\frac{1}{x}$ is smaller than y. In the limit that x becomes infinitely large, $\frac{1}{x}$ approaches zero. Thus, crudely, one divided by infinity equals zero. In the same way, one divided by zero equals infinity, for if in $\frac{1}{x}$ we let x become smaller and smaller, until it approaches zero, $\frac{1}{x}$ becomes larger without limit. (Recall that $1/\frac{1}{2}=2$, $1/\frac{1}{4}=4$, etc.)

There are many more unusual facts about infinity, but we cannot dwell on them here. Measuring the infinite must rank as one of the greatest enterprises of the human intellect, comparable with the most magnificent forms of art or music. Mathematics, 'eternal and perfect' in the words of Lord Bertrand Russell, can be used to build structures more beautiful and satisfying than any sculpture. Yet Cantor's edifice of infinity – 'a paradise from which no one will drive us', as his contemporary David Hilbert was moved to say – took its toll. Grappling with the infinite evidently proved such a disconcerting experience that when the respected mathematician Leopold Kronecker pronounced Cantor's work on set theory as 'mathematically insane', he seems to have a struck a raw nerve. Cantor suffered several nervous breakdowns, and eventually died in a mental hospital in 1918.

Enough has been said to show that the properties of infinite sets or collections are frequently counter-intuitive, and common-sense reasoning may well lead to nonsense. Nevertheless, by discovering these elaborate properties, mathematicians can use infinity without fear, so long as they stick carefully to the rules, however strange those rules may seem. The question is, does it matter? Is infinity relevant, or are these observations just an example of nit-picking by mathematicians for their own amusement? The answer is that infinity does crop up repeatedly in theories of the physical world, and never more so than in connection with spacetime singularities. But before tackling that topic, it is helpful to examine some readily visualizable cases where the infinite has intruded into physics, to the bafflement of its practitioners.

In daily life we think of matter as continuous, but if we subdivide and subdivide, eventually we find that it is, in fact, composed of atoms, apparently discrete entities. It used to be supposed that atoms were like solid, indestructible spheres out of which all things were made. Now it is known that atoms themselves are composite objects, containing electrons orbiting around a nucleus consisting of a collection of protons and neutrons. In recent years, physicists have come to believe that the protons and neutrons – and other subatomic particles that can be produced in high energy collisions – are also composites, consisting of either two or three particles called quarks.

Both quarks and electrons, as well as other electron-like particles, are probably truly elementary, having no constituent parts. How are we to visualize them? To be indestructible they could be infinitely hard, impenetrable spheres. However, such a model runs into grave difficulties. Suppose a spherical quark were struck by an electron, like two billiard balls in collision. When the quark recoils it must move as a whole, without distortion, because being infinitely hard it cannot be squashed. This means that all parts of the quark must move simultaneously. For this to occur, the shock of impact on the surface must be transmitted instantaneously to all parts of the sphere. Such a possibility is, in fact, ruled out by the theory of relativity which, as we shall see in the next chapter, forbids any influence, including shocks, to travel faster than light. Consequently, the 'back' of the quark cannot start to move until at least the time has elapsed for the shock to travel at the speed of light (or slower) through the interior. It follows from this that the sphere must become somewhat flattened if it is to move. But if the sphere is squashy it cannot be elementary and indestructible because we could, in principle, pull it to pieces, given enough energy. Moreover, an elastic sphere needs internal forces to produce its elasticity, and it has proved almost impossible to model such forces.

The remaining possibility is to suppose that the quarks and electrons are indestructible because they have no internal parts anyway. That is, they are pointlike entities, occupying no volume – like the dots of Fig. 12. This idea also suffers from difficulties which will now be examined at length.

Both quarks and electrons carry electric charge. The laws of electricity require that all electric charges repel other charges of the same sign with an inverse square law of force – exactly like Newton's law of gravity, but a repulsion instead of attraction. Suppose we temporarily envisage the quark as a finite sphere filled uniformly with electric charge. Each part of the sphere will electrically repel all the other parts because they carry like charge. Now suppose the sphere is shrunk a little. The electric force, which tries to explode the sphere outwards, gets even more ferocious because the electric charge is being packed still more densely together. Therefore, in order to shrink the sphere we must do a great deal of work against this internal repulsion. The work expended becomes stored in the sphere as electrical energy, and could be recovered by allowing the sphere to expand outwards again, using its force to drive a motor (in principle!).

It is clear that as the sphere is progressively shrunk, more and more energy has to be expended, and the sphere itself accumulates this escalating energy. At this stage the inverse square law of force starts to become important, because a simple calculation shows that the energy of the sphere is proportional to 1/(radius of sphere), or $1/x$ say, where x is the radius of the sphere. This is a familiar symbol, and from the discussion given above it may be seen that as the radius shrinks down to zero, i.e. as the particle becomes point-like, so the energy becomes infinite.

What does infinite energy in the particle mean? According to the theory of relativity, mass and energy are equivalent, which is to say that energy has mass. If, therefore, a point-like charged particle has infinite energy it must also have infinite mass and be infinitely heavy. This nonsensical result – that all subatomic particles should be infinitely heavy – is symptomatic of a disease that afflicts all our theories of subatomic matter in one way or another.

At this level of description, the appearance of infinity, while disturbing, is not disastrous. Far from being infinitely massive, an electron is very light, about 10^{-27} gm. In order to continue calculating with the theory of electrically charged particles it is only necessary to ignore the infinity and replace it instead with the sensible, finite value as actually measured in the laboratory. This can only be an interim measure, for the fact that an infinite quantity arose at all shows that there must be something wrong with the simple type of structure, i.e. a point-like particle, that we have considered. Thus, infinity acts rather like a warning flag, telling us that something is badly awry.

In the 1920s it was discovered that electrons, and other subatomic particles, are really rather more subtle objects than mere scaled-down versions of balls. For example, in many ways it is more accurate to think of them as waves. A new theory was needed to accommodate these peculiar features, a theory now known as quantum mechanics. A complete description of quantum mechanics is beyond the scope of this book, but it is dealt with in depth in my book *Other Worlds*.

One fundamental feature of quantum mechanics is the so-called Heinsenberg uncertainty principle, which permits the law of conservation of energy to be temporarily suspended. This hitherto universal law is a statement of the fact that although energy can be readily converted from one form to another (e.g., heat into electricity, electricity into motion or light, etc.) the total quantity of energy

39

always remains the same. If this were not so we could get something for nothing, creating energy out of nowhere to use at whim.

Inside atoms, this is just what happens. Energy can suddenly appear out of nowhere, so long as it disappears again soon. One may envisage this as energy being 'borrowed' from a limitless bank, subject to the regulation that it be repaid promptly. Just how promptly corresponds to the amount borrowed. The more the energy, the shorter the loan, in proportion. For example, the energy needed to hurl a cricket ball ten metres in the air can only be borrowed for a mere billion-billion-billion-billionth of a second – which doesn't allow it to get very far. That is why we do not notice the Heisenberg energy uncertainty in daily life. Inside the atom, however, it is a different story. There, a little energy will go a long way. An electron can borrow enough to jump right out of the atom and back again, an instability that leads directly to measurable effects.

When electrons move about, they tend to shed some of their energy in the form of photons – particle-like packets of light. Indeed, light is usually created by violent electrical disturbances within atoms. The energy carried by the photons has to be supplied by the electron concerned. However, when the Heisenberg principle is taken into account a new possibility arises. The energy required to create a photon can be borrowed. But because it must be paid back the photon is not really free to fly off into the surroundings. It must stay close to the electron, and disappear again shortly, through being absorbed either by another electron or by the same one. The photon therefore only enjoys a very limited life. How long depends on how energetic it is. A photon energetic enough to be seen can only live for a million-billionth of a second, which enables it to travel a mere hundred-thousandth of a centimetre (or one wavelength) before being 'recaptured'. Still more energetic photons don't even get this far.

At this stage, the question arises as to whether there is any limit to the amount of energy that can be borrowed by the Heisenberg arrangement. This is a question closely connected with the internal structure of the electron and whether or not it is a point-like object. If it is, then a photon can be emitted and reabsorbed at the same point, so it need travel no distance at all. Hence, it can be instantly reabsorbed, i.e. its lifetime can be as short as one pleases. It follows that the energy loan has no known limit, for it can be repaid instantly.

These ideas imply that every electron is surrounded by a cloud of energetic photons. The photons on the periphery of the cloud are not very energetic, but those close in carry considerable energy. As the electron itself is approached, so the energy of the photon cloud rises towards infinity. Because we can never separate an electron from this tenacious retinue of photons, the total energy, hence mass, of the electron will be infinite again, this time because of the photon cloud.

It is by no means trivial to sweep this new infinity away, though. In fact, it wasn't until the late 1940s that physicists knew what to do with it. It turns out that by formulating the theory in a fashion that is properly consistent with the theory of relativity, the infinities may simply be dropped without impairing the predictive power of the theory. Fortunately, as far as electric charges are concerned, the infinite quantities can always be replaced by their finite, measured values.

A theory in which the infinities can be swept away is called renormalizable. Although a renormalizable theory is hardly satisfactory, it is at least better than a non-renormalizable theory. Indeed, if it were not for the fact that our theory of electromagnetic forces is renormalizable, much of twentieth-century physics would have been impossible. However, when it comes to the other forces of nature – gravity and nuclear forces – non-renormalizability is just what seems to happen. Consider, for example, the so-called weak nuclear force, which is responsible for some types of radioactive disintegration. The weak force causes a very short-range and rather feeble interaction between subatomic particles. It operates because a particle called a W can be created temporarily by borrowing Heisenberg energy, and exchanged between, e.g., a neutron and a neutrino. Unlike the emission of a photon by an electron, which merely causes a recoil, this emission of a W by a neutron changes it into a proton, while its absorption by the neutrino converts it into an electron. If, then, a second W is exchanged, the pair revert to their original identities. This sequence of events is depicted in Fig. 16.

The weak force is so much weaker than the electric force, that the result of this minor shuffle ought to be a tiny deviation in the motions of the neutron and neutrino. Instead, theory predicts an infinite disturbance. This time, however, the infinity cannot be tossed aside. This and other infinities crop up again and again in the theory in different places, totally destroying the theory's predictive power. Every time one infinity is displaced, another appears. The same is true

41

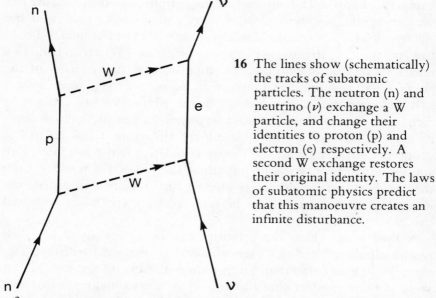

16 The lines show (schematically) the tracks of subatomic particles. The neutron (n) and neutrino (ν) exchange a W particle, and change their identities to proton (p) and electron (e) respectively. A second W exchange restores their original identity. The laws of subatomic physics predict that this manoeuvre creates an infinite disturbance.

of quantum gravity – this theory is also non-renormalizable. An unending string of infinities seem to pollute the theory and render it meaningless.

In the case of the weak force, recent work indicates a way round the problem. By carefully combining the weak and electromagnetic forces into a single theory, it is possible to renormalize the combination, even though the weak piece, on its own, suffers from an uncontrollable attack of the infinities. There are also hopes that the other nuclear forces can be treated similarly. So far, though, nobody knows how to handle the infinities of quantum gravity.

Before leaving the subject of subatomic particles, an important historical episode concerning infinities must be described. When the structure of the atom was first discerned by Lord Rutherford and others in the first decade or two of this century, there seemed to be a profound paradox. It was known that the electrons in an atom revolve round the nucleus like the planets round the sun, bound not by gravity but an electric force of attraction. The simplest atom is that of the element hydrogen, which consists of a single proton orbited by a single electron. As the electron revolves, it accelerates because it is forced to move in a curved path. It has long been known that an accelerating electron emits electromagnetic radiation such as light, so it was natural to suppose that the electron in the hydrogen atom emits light.

The energy of the light must be paid for by the electron, which drops a bit closer to the proton as it loses energy. But the closer it gets, the faster it has to revolve, because of the inverse square law of electric force. This is the same as the reason for the fact that the inner planets orbit the sun faster than the outer planets. Consequently, the electron radiates energy faster and faster, spiralling in towards the proton. As the atom thereby collapses, the energy radiated grows, apparently without limit.

The question now arises as to the structure of the electron and proton. How close can they come together? Forgetting for the moment that the proton is made of quarks (quarks were unknown before the 1960s), we may assume that the electron and proton are structureless, pointlike entities. This means there is no limit to how close the two particles may approach, and hence no limit to the electromagnetic energy that the collapsing hydrogen atom may radiate. To make matters worse, the electron actually reaches the proton, i.e. the separation actually reaches zero, in a finite time – in fact, in far less than a second. So not only is the atom unstable, but it actually threatens to churn out an infinite quantity of radiation in a finite time. This characteristic feature of encountering infinity after a finite, indeed brief, duration will recur later when we examine gravitational collapse. It is a sign that there is something seriously wrong.

In the case of the hydrogen atom, what was wrong was the failure of the physicists of the day to take into account the wavelike aspects of matter on the atomic scale as predicted by the quantum theory. When this was eventually done in the 1920s, it was realized that the electron is actually confined to certain definite orbits, or energy levels, corresponding to a sort of resonance in the wave vibrations. In these levels the electron cannot radiate. Light is only emitted when the electron makes a transition between levels. This frequently happens, and the electron jumps discontinuously down from one level to the next. There is, however, a lowest energy level, or ground state, below which the electron cannot venture, just as there is a lowest note that can be played on a plucked guitar string. Catastrophic collapse does not, therefore, occur. In this case, quantum theory came to the rescue. Whether it can do the same for gravitational collapse is a topic to which we shall return later.

One final point is worth noting. Leaving the wave aspect of the particle aside for the moment, and ignoring the emission of radiation, we can say that if the electron and proton are both released from rest, they will apparently fall straight towards one another under their mutual electric attraction. If they begin one centimetre apart, they will collide after less than a thousandth of a second. However, if there is any initial motion whatever (other than directly towards each other) the two particles will miss each other. The reason is that, as the particles only occupy a single point, they have to line up *exactly* on target to strike. The slightest sideways drift, however small, will prevent collision. What happens then is that the electron flies close to, but straight past the proton, whirls violently around it, and 'bounces' back out again on an elongated elliptical path. When the effects of radiation are included, however, then the electron is disturbed from the elliptical path and can hit the proton target, even if it is a single point.

Facing the infinite has proved a harrowing experience for both mathematicians and physicists, not to mention philosophers and theologians, some of whom were obliged to face death as well. Yet we live in a world that abounds with infinity – in the structure of space and time, the composition of matter, the motion of simple objects and the internal structure of atoms. In the next chapter we shall see that the weirdest case of all occurs when gravity seizes control of matter – and runs off to infinity.

3 Space and time in crisis

> Nothing puzzles me more than time and space; and yet
> nothing puzzles me less, as I never think about them.
> *(Charles Lamb 1775-1834).*

These words of a notable English essayist succinctly express most
people's instinctive reaction to the mention of space or time. They are
things that we take for granted. Their immediacy discourages closer
analysis, for it makes us feel uneasy. Space and time are simply *there* –
an arena in which the world plays out its endless drama – permanent,
dependable and immutable.

When Einstein began to tamper with familiar and cherished beliefs
about space and time, scientists had little choice but to re-examine the
traditional models. Time had usually been regarded as a continuous
flow, like a stream, stretching into the infinite past and future. Above
all, time was uniform and universal, never faltering, never changing;
nature's all-embracing regulator of activity. Aristotle informs us that
'the passage of time's current is everywhere alike'. Changes in material
things, he goes on, 'may be faster or slower; but not so time'. Newton
is also explicit about the absolute and universal nature of time:
'absolute, true and mathematical time . . . flows equably without
relation to anything external'. Above all, so tradition maintained,
temporal durations are independent of material bodies or the location
and behaviour of the observer who measures them.

Space, too, was for centuries regarded as immutable and fixed.
Newton, again, was clear about this: 'absolute space, in its own nature,
without relation to anything external, remains always similar and
immovable'. James Clerk Maxwell, writing about matter and motion

in the nineteenth century, conceives of absolute space 'as remaining always similar to itself and immovable'. For space to move, he reasoned, would amount to a place moving away from itself.

We shall see how seriously in error these early ideas of space and time turned out to be. Not only are these entities not universal and absolute, being in fact subject to change and distortion, but they can move about with a violence exceeding all the other forces of nature.

Things began to go badly with the tidy scheme of a fixed, universal and unchanging space and time about the turn of the century. The central crisis concerned the motion of light signals, a topic that will prove to be of such importance to the subject of this book that the issues will be considered at some length.

To help focus on the peculiarities involved, imagine two observers, A and B, separated by a great distance. They set themselves the task of measuring the speed of light (which is considerable – about 300,000 kilometres per second) by timing the passage of light pulses between them. Suppose that their separation is 300,000 kilometres, and they carefully synchronize their clocks at the start of the experiment. A then sends a light pulse to B at a pre-arranged time, and one second later B observes a flash – the arrival of the pulse. This was the technique used in 1675 by Olaus Roemer, who measured the speed of light by timing its duration of travel across the solar system from Jupiter, which takes up to about an hour. He did not, of course, have the obliging assistant A, but used instead the motion of the Jovian moons, the positions of which could be calculated in advance. The moons always appeared from Earth to arrive late at their calculated positions, due to the fact that the light has to reach us from Jupiter across the intervening space. A measurement of the delay therefore furnished a reasonably accurate value for the speed of light, from knowledge of the distance to Jupiter.

Suppose we now consider the experiment to be a little more sophisticated. The observers A and B wish to check whether the speed of light varies from place to place. To investigate this each of them measures not only the time of light-passage between them, but also the time taken for the same light pulses to travel down a one-metre tube at both locations. The latter measurement requires some fancy electronics as the travel time down the tube is less than a hundred-millionth of a second. After some experimentation, A and B do indeed establish not only that the speed of the light pulses is the same along A's tube as B's tube but, moreover, this speed agrees exactly with the average speed taken over the whole course.

The experiment is now varied a bit. Instead of both A and B remaining at rest, B starts to move towards A. As he is now approaching the incoming light pulses, B naturally expects that these pulses will, on arrival, be found to travel down the tube he is carrying somewhat faster than in the previous experiment. To his astonishment the speed does not change. He asks A to check that the outgoing speed is unshifted and A replies that all is well – nothing has changed at his end. Moreover, the overall speed deduced from timing the delay of the pulses in travelling between A and B shows the same speed as before.

In his consternation, B boards a powerful rocket and thunders off towards A at full power. Faster and faster B races towards A in an attempt to hit the incoming light pulses more rapidly, but the locally measured 'speed-down-the-tube' obstinately remains the same. Soon B has reached half the speed of light himself and he notices that the incoming pulses now look very blue. This is a familiar phenomenon. B realizes that blue light means high-frequency light and remembers that sound waves also shift to higher frequencies when the source and observer rush towards each other. The effect is called the Doppler shift and is noticeable in the higher pitch of a rapidly approaching motor car engine. The sound waves get 'bunched up' in front of the oncoming car.

In spite of the Doppler shift, B registers no change whatever in the *speed* of light. Soon he is travelling 99 per cent of the speed of light straight at the pulses, but still they come at him at only 300,000 kilometres a second. B asks A to make a final check that he is not sending out slow light pulses, but A replies that his pulses are all leaving at full speed – 300,000 kilometres a second. The two experimenters begin to realize that something very strange is happening.

As a double-check on these remarkable results, B turns the rocket about and blasts away at full power in the opposite direction, running fast before the light pulses. He notices that the light now looks very red – the waves are being stretched out like the drop in pitch of a car engine as it rushes past and then recedes. Soon B is receding from A at 99 per cent of the speed of light. He expected the light to be only just overtaking him, at only one per cent of 'normal' speed (i.e., at only 3000 kilometres per second). But nothing of the sort! The pulses still keep coming up behind at the same speed as they had when B was rushing towards them. His colossal change in motion, totalling almost twice the speed of light, has not changed the speed of the pulses by even one kilometre per hour. B checks one last time with A that he is not now shooting the pulses harder, but A measures exactly the same pulse speed as B, even though they are in rapid relative motion.

A final desperate attempt is made to break the deadlock. B announces he will switch on the special rocket overdrive and outrun the pulses altogether. Obviously by travelling faster than light, the pulses will not even be able to reach him. At some stage between 99 per cent and 101 per cent light speed, therefore, the pulses *must* slow up and fall behind. While this part of the experiment is in progress, A notices that B is slowly nudging up towards the speed of light, but the nearer he gets, the more power he needs to accelerate. The power requirements seem to grow without limit. Even with all the energy in the world, B cannot gain that last little bit needed to break the light barrier. It seems as though the closer B gets to light speed, the heavier the rocket becomes. He needs more and more power just to accelerate a tiny amount. The extra energy all seems to go into creating more and more mass – not more speed. And still the pulses come on at the same speed as when the experiment started. Eventually, with fuel depleted, the experimenters abandon all attempts to alter the locally measured speed of light from 300,000 km per second.

This little story is a modern version of experiments actually performed in the late nineteenth century and repeated, in one form or another, many times since then. It highlights the almost paradoxical nature of light propagation, for both A and B measure the same speed of light, even though they are in rapid motion relative to each other. The explanation was provided by Einstein in his so-called special theory of relativity, published in 1905. Einstein proposed that the speed of light is always the same, for everybody, no matter what their state of motion, or however the source of light moves. To make sense of this restriction it is necessary to assume that *space* and *time* rather than the speed of light change with motion. For example, when B rushes towards A, the distance from A to B, as measured by B, shrinks. If B observes familiar objects surrounding A they all look very flattened. Of course, by A's reckoning these objects appear to be quite normal.

In addition to the peculiar shrinkage of space, B's motion has a weird effect on time. B regards A's clock as running slow relative to his, while A sees B's clock running slow. Not only are their images of spatial distances discrepant, but so are their time scales. Indeed, when A and B finally get together at the end of the experiment, B finds that his watch is many hours behind A's clock, due to all the rushing about. The sequence of events between the start and completion of the experiment has occupied much more time for A than the *same* sequence of events has occupied for B.

What extraordinary ideas these are; they fly completely in the face of the long-established picture of space and time quoted at the beginning of this chapter. Since Einstein's theory became accepted, and experiment after experiment has verified its predictions, the distortions of space and time caused by motion have never ceased to be a novelty. Popular science books dwell at length on the oddities of ultra-rapid space travel, involving astronauts returning from short trips round the cosmos to find a planet that has moved on thousands of years. What makes these images so intriguing is the demise of the common-sense picture of the world where space and time – distance and duration – are fixed for all observers, whatever their state of motion. Einstein's theory fixes instead the speed of light, leaving space and time to alter themselves around the observer's motion in such a way that this speed is always the same.

From the foregoing it is clear that no amount of acceleration will enable a rocket, or any other material body, to exceed the speed of light. Space and time will go on distorting indefinitely so that the light barrier remains unbroken. In modern subatomic particle accelerators, fragments of atoms are whirled to within 0.01 per cent of the speed of light. It is found they grow dozens of times heavier than when at rest, making it ever more expensive to increase their speed still more. And not only matter, but even signals cannot travel from place to place faster than light. This demolishes the almost unanimous assumption by science fiction writers that space age communication can take place instantaneously across the universe.

For a better understanding of Einstein's theory of space and time a diagram can be drawn. Figure 17 is called a spacetime map, and shows time running vertically, with one dimension of space running horizontally. There is nothing complicated or mysterious about a spacetime map; it is simply a graph of where objects are located at various moments. For example, the line marked A represents the perambulations of a body that moves to the right and then to the left. The line marked B represents a particle that remains at rest relative to the

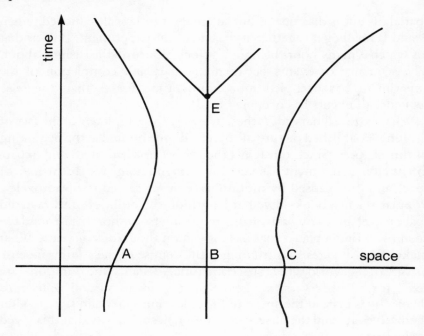

17 Spacetime diagram. Time is measured vertically, space horizontally. The lines may be regarded as the tracks of particles through spacetime, or simply as a graph showing where the particles are at successive moments. A point such as E is called an event. The light rays from this event travel out through space at constant speed, so they are drawn as straight lines on our diagram.

person who has drawn the diagram. Line C represents another particle that first oscillates back and forth, then shoots off to the right at ever-increasing speed. These tracks on the space-time map are called 'world lines' and show the histories of moving bodies.

Light pulses may also be drawn on the diagram. If we measure time in seconds and distance in light seconds, then the paths of light pulses are straight lines at 45°. Two oppositely directed light pulses are shown leaving body B at some instant marked E.

If an observer is not at rest relative to body B, but travelling fast to the left, B will appear to him to be moving to the right. The other bodies too will have, superimposed on their motion, an extra right-wards velocity. If this new observer draws a spacetime diagram it will contain roughly the same features as Fig. 17, but the world lines of the bodies will be sloped over towards the right somewhat. However, notice that the light rays must still be at 45°. They are not affected by the new perspective because the moving observer still sees the light pulses travelling at 300,000 km per second. Thus, the light lines on the diagram are a fixed and invariant feature, independent of the motion of the observer.

The fact that the light lines remain the same for all observers irrespective of their motion has a far-reaching geometrical significance first spotted by the German physicist Hermann Minkowski in 1908. The spacetime diagram viewed from the perspective of a differently-moving observer will have the regions away from the light lines rearranged because of the motion. The rearrangement, however, is highly restricted so that the light ray paths remain unaltered. It follows that space and time must adjust themselves in co-operation to avoid upsetting these rays. It has been mentioned that distance and time intervals are both distorted by motion. We see that the changes in each are arranged in a way that will leave the light ray paths unchanged. The conclusion is that space and time are not independent entities, but linked together in a very intimate way so that light may enjoy its special properties. For this reason, physicists no longer think of space and time, but of spacetime – a unified four-dimensional structure.

Another dimension can be added to this scheme as shown in Fig. 18. If we imagine taking horizontal slices through the three-dimensional drawing, then each slice represents space at one moment of time. A higher slice in the stack represents space at a later time. The world lines can now spiral about, representing motion in more than one dimension. For example, the helix shown in the diagram is the world line of a body that revolves round and round in a circle.

If a pulse of light is suddenly emitted in all directions by a stationary body, then the paths of the outgoing rays lie along a cone in this three-dimensional picture. The cone is generated from the oblique light lines in Fig. 17, rotated about a vertical axis. This structure, called the *light cone,* will play a central role in the topics to follow.

Of course, the two-dimensional slices that represent space at one moment are only an imperfect image of the real world, where space is three dimensional. The diagram attempts to show both space and time, and if we were to use all directions on the diagram to represent the three dimensions of space, none would be left over for represent-

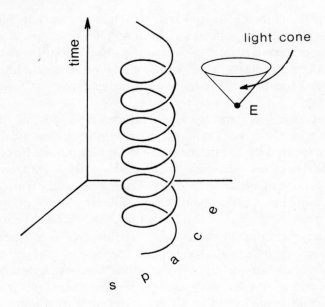

18 This spacetime diagram shows two dimensions of space. Horizontal slices represent all of space at one instant. The helix is the world line of a particle that moves around in a circle. The cone is the paths of all the light rays that leave event E and travel outwards at constant speed in all directions. The circles obtained by slicing the surface of the cone horizontally must be envisaged as the sphere of light emitted by E, expanding outwards.

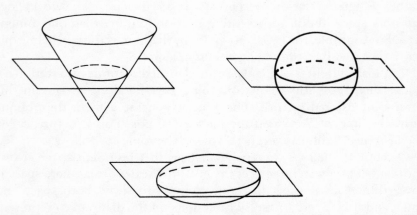

19 Three-dimensional objects when sliced give two-dimensional sections.

ing time. So in what follows we make no attempt to represent fully all of space and time (which is *four* dimensional) and rely on the reader's ability to handle a two-dimensional representation of three-dimensional space.

To be completely clear about this, Fig. 19 shows a number of shapes together with their two-dimensional representations. One can think of the two-dimensional version as a slice or section through the solid object. For example, the sphere becomes a circle, which would be obtained by slicing anywhere through the sphere.

The horizontal sections through the spacetime diagram can be considered as snapshots of the universe at one moment. If the slices are imagined as stacked up on top of each other, the snapshots become a movie film, and we can follow, frame by frame, how a system evolves. For example, the horizontal slices through the light cone are circles of ever-increasing radius. In accordance with the above-mentioned correspondence between circles in the two-dimensional space sections and spheres in real three-dimensional space, one deduces that these 'snapshots' of the light cone represent spherical surfaces of light. These spheres are images at successive moments of the wavefront of an omnidirectional light pulse, as it spreads outwards in all directions. Slicing higher and higher up the diagram, the circles grow larger, representing the expanding sphere of light in real space. The light cone is of central significance in any analysis of the structure of spacetime because of its relation to causality. As nothing can travel faster than light, no causal influence from its apex can ever cross to the exterior of the light cone.

A point in spacetime is a particular place at a particular moment. Physicists call such a point an event, and think of the whole of spacetime as an infinite collection of events. Of course, an event in this sense does not imply that anything especially spectacular is happening, but if we think of all points of space as having a tiny imaginary clock, then an event is just an instant on one such clock. The basic issue of causality then concerns which events can influence other events. Before Einstein, it was supposed that any two events could, in principle, be connected by a signal or influence of some sort. Gravity, for example, was supposed to propagate instantaneously through space, so that if the moon were suddenly to disappear miraculously, the tides would stop at the same instant, even though we should not *see* the disappearance of the moon for about another second, this being the time required for moonlight to reach the Earth. After Einstein's special theory of relativity it was realized that no signal could travel faster than light without producing causal chaos.

An easy way of understanding the difficulty is to imagine a rocket travelling at 99 per cent of the speed of light, passing the Earth (see Fig. 20). At the exact centre of the rocket is a mechanism which sends a

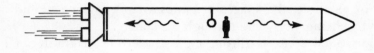

20 From the rocket, the two light pulses, travelling at equal speed, strike the end bulkheads simultaneously. From the Earth, however, the pulses *also* appear to travel at the same speed, so the left-hand pulse arrives at the oncoming bulkhead first.

short pulse of light in both directions down the length of the rocket. Naturally, to an observer in the rocket the pulses will appear to arrive at the end bulkheads simultaneously, because both pulses travel at the same speed – the speed of light. However, the situation as witnessed from Earth is quite different. According to the theory of relativity, the speed of light has the same value when measured by the Earth observer. In particular, the two pulses travel at the same speed *relative to Earth*. Thus, because the Earth observer sees the rocket plunging forward at 99 per cent of the speed of light, it is clear that, as witnessed from the Earth, the pulses do not arrive simultaneously. Instead, the Earthman sees the rear bulkhead of the rocket advance rapidly to meet its pulse, whereas the forward bulkhead is retreating from its pulse. During the time taken for the light to travel down the length of the rocket, the rocket itself will have moved forward appreciably, and the rear pulse will strike its bulkhead well before the other pulse reaches the front of the rocket. What appear to the rocket observer as simultaneous events seem from Earth to be events separated in time. The conclusion is that simultaneity is relative to one's state of motion. There is no universal, absolute agreement on what the 'same moment' is in two separate locations.

The problems that arise when simultaneous events are no longer regarded as simultaneous by someone else is that there can be no agreed definition of what is 'instantaneous'. A signal which travels 'instantaneously' from the front to the rear of the rocket (e.g. 'the light pulse has arrived'), as judged from within the rocket, would be regarded by an Earthbound observer as a signal propagating *backwards* in time. Because from Earth the front bulkhead is seen to be hit by its light pulse *after* the rear bulkhead is hit, the apparently 'instantaneous' signal from front to rear seems from Earth to be a signal from a later to an earlier event.

The paradoxes that can arise when signals are allowed to travel backwards in time are well known. Consider, for example, a machine with the following instructions programmed into its computer. At three o'clock it emits a signal into the past. The signal is reflected from some distant place and arrives back at the machine at one o'clock. On receipt of this signal, the machine's programme instructs the machine to blow itself apart at two o'clock. This sequence of events is therefore inconsistent because at two o'clock self-destruction would pre-empt the three o'clock transmission, prevent the one o'clock reception, and hence *not* trigger the self-destruct mechanism, in contradiction to the original assumption.

Since faster-than-light signalling is ruled out for causal reasons, it is clear that certain events can never influence, or be influenced by, some other events. This is illustrated in Fig. 21, which shows the light cone emanating from an event E. Also shown is the backward light cone, stretching into the past. This represents a spherical sheet of light converging on to the event from distant space. According to the principle that no signal can exceed the speed of light, events such as E', outside the light cone at E, cannot in any way be influenced by, or themselves influence, E. In contrast, events (such as E″) inside or on

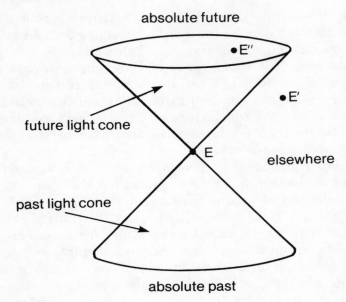

absolute future

• E''

• E'

future light cone

E

elsewhere

past light cone

absolute past

21 The light regulates the causal structure of spacetime. It divides events that can influence one another from causally disconnected events.

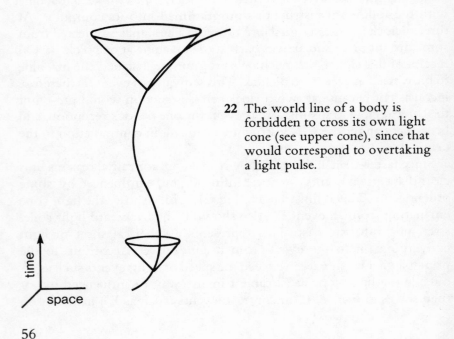

22 The world line of a body is forbidden to cross its own light cone (see upper cone), since that would correspond to overtaking a light pulse.

time

space

the future light cone can be influenced by what happens at E. Also, events inside and on the past light cone can influence what takes place at E. For this reason these regions are labelled the absolute future, absolute past and 'elsewhere'. Such causal relations between events are, as we shall see, a vital ingredient in exploring for the existence of naked singularities.

The fact that all material bodies must travel slower than light means that all along the world line of a body, the light cone fans out around it. The situation shown in Fig. 22, where a world line crosses its own light cone, is strictly forbidden, for it would correspond to the body overtaking a pulse of light. Naturally the world line of one particle can cross the light cone of another. Figure 23 shows how the light cones

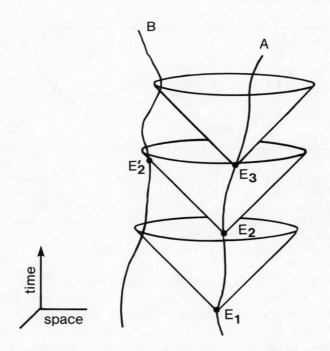

23 The light cones emanating from the world line of body A carry the first news that can ever be known about the successive events E_1, E_2, E_3, ... that befall A. When these cones intersect B's world line he can know of these events. For example, at event E'_2 B learns about A's earlier experience at E_2. B cannot know yet about E_3, but he has already learned about E_1.

emanating from successively later moments along one world line cut a neighbouring world line, bringing information about the events E_1, E_2, E_3, ... No knowledge of these events could be obtained by the other body before these light cones intersect it.

In addition to requiring all bodies to travel slower than light, there is another requirement that is necessary if causality is not to be violated. This concerns the topology of time. To avoid complications, it is necessary that the absolute future and absolute past are distinct; in other words, no event belongs to *both* the absolute future and absolute past of any other event. Clearly this cannot occur in the spacetime diagrams as drawn in the figures so far, but if instead of drawing the diagrams on a flat sheet of paper, the paper is cut top and bottom somewhere, and rolled over to make a cylinder as shown in Fig. 24,

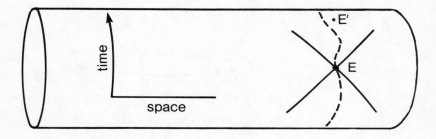

24 Closed time. If spacetime were like the surface of this cylinder, the past would also be the future, and time would be limited in duration. An event such as E′ lies within the future light cone of E, but the past cone, if extended backwards right around the cylinder, would also enclose E′. (To depict this arrangement, only one space dimension has been drawn, so 'cone' means two V-shaped lines.)

then one can clearly expect trouble. Rolling the diagram up to form a cylinder in no way affects the geometry of the lines drawn (all distances and angles are unchanged, for we do not have to stretch, twist or shrink the paper) so it is locally identical to the flat sheet. Nevertheless its global topology is drastically different. Instead of time stretching infinitely into the past and future, time is finite, and equal in length to the circumference of the cylinder. This model of space and time represents a sort of cyclic universe.

The nature of the light cones in such a model universe is very strange. The broken line represents the world line of a body, and it has an event E marked on it, together with forward and backward light cones. The Event E' which occurs near this body lies within the absolute future of E, inside the forward cone. But it clearly also lies in the absolute past of E, within the backward cone. In this spacetime the past can also be the future. Indeed, if the body is an observer, then he will be able to visit his own past by waiting long enough: the world line is closed into a loop, and the forward line cone, if slid forwards along the world line, will, after going once round the cylinder, intersect the past light cone. The spectre of past and future all muddled up in this way is usually enough to suggest to physicists that such a spacetime cannot correspond to reality.

The behaviour of the light cones becomes much more interesting when gravity is taken into account. The effect of gravity on light has been known since Einstein formulated his so-called general theory of relativity, published in 1915. There are two important ways in which gravity affects light. One is the bending of light beams by a gravita-

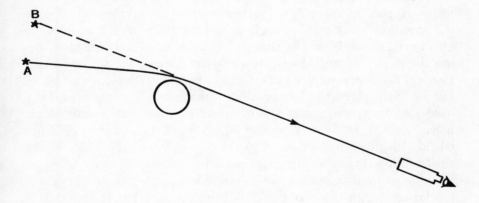

25 Gravity bends starbeams. A star whose position is really A appears to lie at position B when the sun intervenes. The effect is grossly exaggerated in the figure. The observation must be performed during a solar eclipse to cut out the sun's glare.

tional field, illustrated in Fig. 25. A starbeam that passes close to the sun is deflected or bent towards the sun in the fashion shown. As shown in chapter 1, the existence of the effect can be checked by carefully measuring the position of a star when the sun is well away from the line of sight, and watching how this position becomes displaced as the sun moves towards the location of the star in the sky during its annual migration through the constellations of the zodiac. These days it can also be checked using radar rather than light (all electromagnetic waves travel with the speed of light) by bouncing radar pulses off other planets in the solar system when they lie on the remote side of the sun.

Another way in which light is affected by gravity concerns its frequency. As light climbs from the ground upwards it 'tires', i.e. loses energy, just as a material body does. This gravitational depletion is manifested by a loss of frequency, in other words a shift in the colour of the light towards the red end of the spectrum. Another type of red shift – the Doppler shift caused by the recession of a light source – has already been mentioned. The gravitational redshift can be measured directly in the laboratory by sending very finely tuned gamma rays up a tower. The effect, though minute on the Earth, is nevertheless measurable.

An alternative way of regarding these effects of gravity on light is in terms of space and time. The loss in frequency of light is equivalent to the speeding up of time at high altitudes relative to the ground, an effect that can be verified by flying clocks aboard rockets and monitoring their rate as gauged against a master clock at the surface of the Earth. This experiment was performed by R.F.C. Vessot and M.W. Levine of the Harvard College and Smithsonian Observatories. They used a hydrogen maser as a very stable and accurate atomic clock, placed on board a Scout D rocket which was launched from Wallops Island, Virginia, on 18 June 1976. They monitored the 'ticks' of the rocket-borne maser relative to two ground-based masers, using an accurate ground-to-rocket radio link. As expected, the rocket clock ran faster at high altitudes than its identical counterparts on Earth.

It is important to realize that an observer on board a rocket would not *feel* that time was running faster, for his mental and physical processes would be equally affected. For him, events on the surface of the Earth seem to be running slow. In the case of Earth, the effect is far too small for a human being to notice, but we shall see that there do exist extraordinary astronomical objects for which the temporal dislocation between the surface and a distant place becomes enormous.

How can these ideas be depicted on a spacetime diagram? The bending of a light ray by gravity will be represented on the spacetime diagram by a distortion in the shapes of the light cones. In chapter 1 it was explained how, according to Einstein, gravity must be regarded, not as a force, but as a distortion in the geometry of space and time. In a bent spacetime, it is no surprise if the paths of light rays are bent (as in the case of the deflection of a starbeam by the sun) or stretched (as in the red shift experiment mentioned above). We can envisage this distortion of the light cones as due to the bumps in spacetime caused by the presence of gravity.

Figure 26 (i) shows several light cones in the absence of gravity. Figure 26 (ii) shows the same cones when a large gravitating object A, such as a star, is present to the left (its world line is marked with a heavy vertical line). As can be seen, the effect of the gravity is to tilt the cones leftwards somewhat. The curved line is the world line of an observer who falls freely downwards in the gravitational field.

Sometimes it is helpful to view these diagrams from above. Each cone then becomes a circle with a dot to represent its apex. From this perspective, the effect of the gravitational field is such as to displace the circles somewhat towards the gravitating centre, as though the light is being dragged over towards it. These pictures are entirely equivalent to the ones that would be drawn if, instead of using the flat sheet of this book, a curved and distorted sheet of paper were used to represent curved spacetime.

Armed with these pictures, let us examine the effect of gravity around a massive star. Figure 26 (ii) depicts this situation in two ways: by showing the light cones from the side and viewed from above. Near the surface of the star, gravity is very strong and the light cones are tipped far over, but farther away the effect is less severe. The strength of gravity, it will be recalled, rises in accordance with the inverse square law as the surface is approached.

It might be wondered what happens if the star is further shrunken so that the gravity near its surface rises even more. How far can the light cones tip? It is possible to calculate the amount of tip for a spherical body, and these calculations show that a star with a mass equal to our sun, when shrunken to about 1.5 km in radius, produces a bizarre effect in the surrounding spacetime. This effect is depicted in Fig. 27. The light cones have tipped so far over that the right–hand surface of the future cone actually crosses the vertical and slopes inwards (i.e. leftwards) towards the star. Figure 28 shows the same phenomenon as viewed from above. Inside a critical radius, called the Schwarzschild radius after the German mathematician Karl Schwarzschild who first

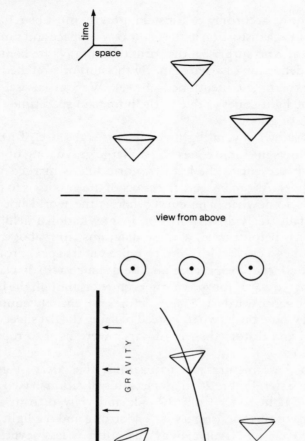

(i)

view from above

GRAVITY

view from above

(ii)

A

26 When a massive object A (such as a star) is present, its gravity drags light towards it. The effect on our spacetime diagram is to tip the light cones over in the direction of A. The cones nearest A are most affected. Viewing the cones from above one sees the edge of the cone as a circle and its apex as a dot. In (i) there is no gravity and the dots lie at the centre of the circles (upright cones). In (ii) the tilted cones cause a displacement of the circles. This distortion is equivalent to a bent spacetime.

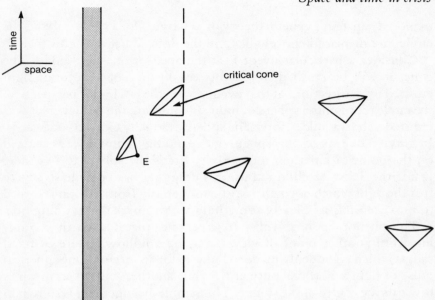

27 If the gravity near a spherical star (drawn as a hatched tube) becomes strong enough, it can tip the light cones right over, so that they point inwards (towards the star) rather than upwards. The broken line is the critical Schwarzschild radius, where the right edge of the cone is tipped to stand vertical. An event such as E cannot be seen from outside the Schwarzschild radius.

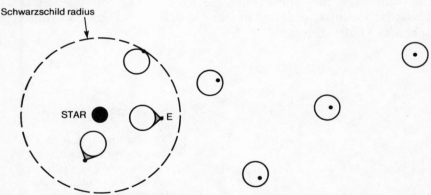

28 The situation depicted in Fig. 27 is viewed from above. The sides of the cones remote from the star try to carry light away from the gravitating centre, but inside the Schwarzschild radius the dragging effect of gravity is so strong that even this outward edge of the cone slopes inwards, sweeping the light towards the star. Thus an event such as E cannot send information outwards.

explored Einstein's general theory of relativity for a spherical star, the circles are displaced entirely beyond the dots. What does this mean?

Consider a particular event E at the apex of one light cone. The cone, it will be recalled, represents the history of a pulse of light emitted in all directions at that event. From the circle–dot perspective, the circle represents a sphere of light surrounding the dot where it was emitted a short while before. The sphere gradually expands outwards. In gravity-free space, this sphere of expanding light is always centred on the point of emission, but gravity drags the sphere to one side. Inside the Schwarzschild radius, this dragging has become so severe that the light which normally would move *away* from the star is forced to move *towards* it. Clearly something very remarkable is going on.

This phenomenon is so alien to experience that it is worth examining from another point of view. Imagine a hollow sphere of fixed radius which is suddenly made to light up for an instant. One spherical pulse of light will travel outwards, while another will travel inwards towards the centre of the sphere. This simple arrangement is shown in Fig. 29 (i) where the light spheres a short while after emission are shown as broken lines. If the hollow sphere were located around the compact star, inside the Schwarzschild radius, the behaviour of the spheres of light would be startlingly different. The ingoing, contracting sphere would shrink down towards the centre as before. However, this time even the *outward*-directed light would converge inwards towards the centre, as shown in Fig. 29 (ii). Both outgoing and ingoing spheres of light shrink.

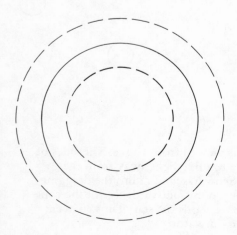

29 Pulses of light in a hollow sphere:
(i) The hollow sphere (solid line) lights up for an instant. A short while later the spherical pulses of light emitted have reached the positions shown by the broken lines. The outgoing sphere of light expands, the ingoing one shrinks towards the centre.

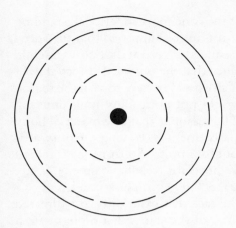

(ii) When the hollow sphere is enclosed within the Schwarzschild radius of a massive body (blob), *both* spheres of light *shrink;* even the outward-directed pulse travels inwards. Consequently, the hollow sphere could never be seen from outside.

Using physical language, the extraordinary behaviour of the outward-directed light is sometimes described by saying that gravity near such a shrunken star has become so intense that it snatches back even light – the swiftest entity in the universe. The possibility that the universe might contain stars so massive that their gravity even traps their own light was conjectured nearly two hundred years ago. It is helpful to envisage this exotic situation using an analogy.

Suppose a pulse of light is represented by a ball and the effect of gravity by a string of elastic fixed to a post, such as one occasionally encounters in certain types of children's games. If the string is stretched and the ball is thrown towards the post, the effect of the elastic tension is to increase the velocity of the ball. But if the ball is thrown away from the post, the elastic force impedes its progress, reducing the velocity of projection. If the string is shortened and stretched to the same place, the tension will be even greater, and a critical point will be reached where the outward-thrown ball will be overwhelmed by the elastic force and instead of moving away from the post, it will be snatched backwards out of the thrower's hand. This corresponds to the situation inside the Schwarzschild radius, where gravity snatches back outward-directed light.

If the hollow sphere is located exactly at the Schwarzschild radius, the outward-directed light rays remain stationary – they travel neither outwards nor inwards, but hover at a fixed radius from the central object, hugging the surface of the hollow sphere. In Fig. 27 this

65

corresponds to the outside edge of the cone marked 'critical' standing exactly vertical. Nevertheless, it must not be supposed that the speed of light in the usual sense has been reduced to zero at the Schwarzschild radius. It is only from a great distance that the outward-directed light rays appear to make no progress. If an observer were to visit the region of the Schwarzschild radius he would not notice anything unusual, such as sluggishly moving light. The reason is that any material body in this region will be plunging rapidly inwards, leaving the outward-directed light to struggle its way out to infinity. So from the frame of reference of the falling observer, light seems to behave normally, i.e. to travel relative to him at the usual 300,000 km per second.

One obvious consequence of the fact that even outward-directed light gets dragged towards the centre of the star is that no light at all can reach the outside world from inside the Schwarzschild radius. Therefore, because no information can leave an event faster than light, the outside world can never have knowledge of events that occur inside the Schwarzschild radius. The spherical surface at this radius therefore divides spacetime into two distinct regions: the exterior events that we can, if we wait long enough, witness, and the interior events that are forever hidden. The spherical surface at the Schwarzschild radius is therefore called an *event horizon*. Just as with a terrestrial horizon, its existence does not imply that nothing exists beyond it – events there can be witnessed by travelling across the horizon. But an observer who remains for ever outside the event horizon can have no knowledge about the interior region. The horizon itself can be thought of as marked by the last light pulse that can reach a distant place after a very long time. It is formed from the outside edge of the light cones that are just tipped exactly vertical in Fig. 27. Because the region inside the horizon cannot be seen, an object like this would appear completely black, and so it is called a black hole. Black holes are thought by astronomers to exist in many parts of the universe, and their properties are explained in detail in my book *The Runaway Universe*.

Although the above discussion was given for the special case of a spherical star, event horizons form under far more general circumstances. In particular if the star is not quite spherical, and perhaps rotating, the major details of black hole formation remain the same. When a star collapses through the event horizon the gravitational red shift discussed earlier in this chapter becomes very important. The light which leaves the surface of the star has to climb against a gravitational field that is becoming ever stronger as the star contracts. On reaching a distant observer, this light becomes, therefore, redder and redder. As the surface of the star approaches the critical radius where it is about to cross the event horizon, this red shift escalates without limit. The colour of the star becomes very dull and eventually goes black as the object converts to a black hole. In a typical star this fade-out occurs exceedingly rapidly – in a few millionths of a second.

According to the picture, already discussed, that the red shift of light can be viewed as a slowing down of time at the star's surface relative to a distant place, the fact that the shift grows without limit implies that time on the surface of the collapsing star literally grinds to a halt. As far as the distant observer is concerned, events on the star become frozen in time, and all activity is suspended. He cannot see this immobility, though, because the star has turned black. In contrast, events as witnessed on the star itself are quite different. An observer stationed there would not notice any slowing down of time. In the millionth of a second (in his frame of reference) that it takes for him to reach the event horizon, all of eternity will have passed outside.

Once inside the event horizon, what happens to the star? At first sight it seems that the progressive crushing of the stellar material must eventually produce a lump of matter so compressed – so hard and dense – that the shrinkage will stop and the object will remain at a fixed, though admittedly tiny, radius. However, there is a serious objection to this conjecture. The surface of the star, being made of ordinary matter, cannot exceed the speed of light and must therefore be forced to remain inside the light cones. But as can be seen immediately from Fig. 27, the light cones inside the event horizon are tipped over severely. Thus the world line of the star's surface is forced inwards with the light. Also in Fig. 29 (ii) the stellar surface *must* lie between the two light spheres (broken lines). But *both* of these spheres are contracting, so the star's surface must likewise contract. It is trapped between the two shrinking spheres as though in the jaws of an infinitely powerful vice, and so cannot avoid being progressively more crushed. (For this reason the hollow sphere in Fig. 29 could not be made of ordinary matter, or it would collapse. It must be envisaged as an imaginary spherical surface rather than a material shell.)

The more the star shrinks, the stronger this crushing becomes – the light cones are tipped ever more inwards. The star itself escalates in its

67

rate of shrinkage. No force in the universe, however strong, can support the star against its own immense weight. The star's gravity here overwhelms all else and forces the star with increasing rapidity into an ever-decreasing volume of space.

So what happens? If the object is exactly spherical, the progressive, cataclysmic shrinkage will squeeze the entire star into a single, mathematical point at the centre. (So we should really think of the central object or 'star' in Figs 27–29 as a point rather than a spherical blob.) If this happens, then the density of material rises without limit and becomes infinite. This already indicates that something has gone badly wrong. When infinity is predicted for a measurable physical quantity, only two alternatives present themselves. Either the theory has collapsed, or the physical world has ceased.

In the previous chapter, a number of other examples were given in which physicists had discovered infinity in their equations. In the case of the collapsing atom, the old theory had to be replaced by the new quantum theory in order to recover sensible predictions. Here the task is not so easy. A halt to the collapse would imply that the material of the star is able to exceed the speed of light, thereby opening up the prospect of causal anarchy inside the black hole, involving travel into the past, with all its attendant paradoxes.

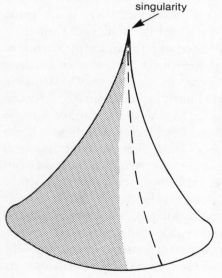

singularity

30 Spacetime (the surface) curves progressively more until it pinches off altogether at a point, and stops. A curious observer (broken line) who explores near the tip risks disappearing for good off the end – he can never return. The singularity at the tip represents a limit to space and time.

If the inexorable crushing does not stop, then more is at stake than the fate of the star. Einstein's theory connects gravity with the geometry of space and time. Roughly speaking, the stronger that gravity grows, the more distorted or curved spacetime becomes. In the case of a spherical star shrunk to a single point, the gravity at the surface of the star rises to infinity (see Fig. 1) which implies that the curvature of spacetime also becomes infinite there. Physicists call such a phenomenon a spacetime singularity. What does it mean physically?

One way of envisaging a spacetime singularity is by analogy with the curved surface of a cone-like structure, as shown in Fig. 30. Near the base of the cone, the curvature of the surface is rather shallow, but as one passes up the stem towards the apex, the surface becomes more and more curved, until at the tip of the cone the curvature becomes infinitely great (a singularity). The cone cannot be extended past this point. A path that leads up towards the apex (broken line) cannot continue upwards indefinitely. Once it hits the singularity, the surface stops. In analogy, a world line in spacetime, when it runs into a singularity, cannot return – spacetime 'stops'.

There can surely be few predictions in science as bizarre as this. The end of spacetime signals the end of the physical universe, at least as we understand it. The full implications of an abrupt cessation to all existence will be explored in chapter 7. Here we simply note that if the collapsing star encounters the spacetime singularity it must run off the edge of the physical world, never to return. In that case the object that implodes to form the black hole suddenly ceases to exist after the fleeting duration of collapse. So unpalatable is this conclusion, that physicists have long been reluctant to believe it.

One escape route might be to question the assumption of precise sphericity. Obviously no star can be *exactly* spherical, and without this symmetry it is not clear that all the star will shrink to the same point. For a better understanding of this qualification, imagine the star to consist of a ball of billions of point-like particles. As the ball collapses, each particle falls precisely radially, towards the exact centre of the ball. Obviously, because of this high degree of symmetry, all the particles will arive at the same place, i.e. the centre, and the density will become infinite. If, on the other hand, the system were not precisely spherical, then the particles would fall towards slightly different places and so 'miss' each other. But where do they go? It seems that they must plunge right on through and out the other side, so the collapsing ball bounces at a very small radius, then expands out again. However, it cannot expand back out into our universe, because it is trapped inside the black hole by the event horizon. The event horizon represents, it will be recalled, a surface in space at which, from the point of view of a distant observer, time stands still. Thus, the implosion of the star, while taking only a microsecond star-time (i.e., in the star's frame of reference), takes an infinite amount of time as viewed by the distant observer. Consequently, if the collapsing star were to bounce back out of the black hole into our universe, then it would be seen to emerge before it had even collapsed through the event horizon, which is absurd.

Three alternatives have been suggested. The first is that the star bounces out into another part of the universe, bubbling into space beyond some distant galaxy perhaps. This idea seems to run the risk of temporal paradoxes too, and is not taken very seriously. The second is that the star emerges into a completely different universe – another space and time co-existing in parallel with our own and unconnected with our universe until the star collapses. Thirdly, and most disturbingly, is that in spite of the more complicated shape of a non-spherical collapsing star, nevertheless all or part of the star will still hit a singularity and cease to exist.

The star could still reach infinite density even if it does not shrink to a single point. For example, it could become very flattened – like a pancake – and shrink to zero thickness. Alternatively, it could become drawn out into a thin cigar shape which would then shrink to a line of infinite density and zero thickness. In a realistic collapse, it is clear that even these shapes are highly simplified, and it may be that much more complicated shapes occur which are still infinitely dense.

About fifteen years ago many physicists believed that in a real collapsing star, departures from spherical symmetry would enable the star to avoid reaching a condition of infinite density. It was further believed that, under these circumstances, spacetime would also avoid becoming singular. Then on to the scene came Roger Penrose, a brilliant mathematical physicist then working at Birkbeck College, University of London. In 1965 Penrose proved a result with stunning implications, and opened up a whole new chapter of physics. He showed that even if the collapsing star avoids reaching a condition of infinite density, nevertheless spacetime itself cannot avoid becoming singular in some way or other. The Penrose theorem, and the related work that followed it, herald a dire and apparently inescapable crisis for all of nature.

4 Towards the edge of infinity

The discovery of spacetime singularities was a pivotal event in the history of science. In 1915 Einstein liberated space and time from the requirements of immobile rigidity imposed by Aristotle and Newton. As a result, physicists conceived of a changing, moving spacetime – a geometrical arrangement that could evolve and transform itself in an infinite variety of ways. But the price exacted for endowing spacetime with life and activity is that, once gravity runs out of control, spacetime smashes itself out of existence at a singularity.

Whether or not we understand fully just what a singularity is, it is vital to know under what circumstances one will occur. The importance of Penrose's theorem about singularities is that it demonstrates that they are not merely mathematical quirks caused by overidealization. The theorem indicates that singularities, far from representing a very special sort of end to gravitational collapse, are almost inescapable features of a gravity-powered cosmos.

The theorem itself hinges on the existence of the bizarre phenomenon discussed in the previous chapter, namely, that inside a black hole gravity is so strong that even the outward-directed light rays get dragged inwards. Penrose considered this phenomenon quite generally, without needing to assume that there exists an exactly spherical collapsing star.

Suppose that because of extreme gravity in some region of spacetime, the light rays suffer this inward dragging. One can imagine a surface surrounding this region – not necessarily a spherical surface as considered in the previous chapter, but it should at least have the same topology as a sphere – from which outward-directed light is

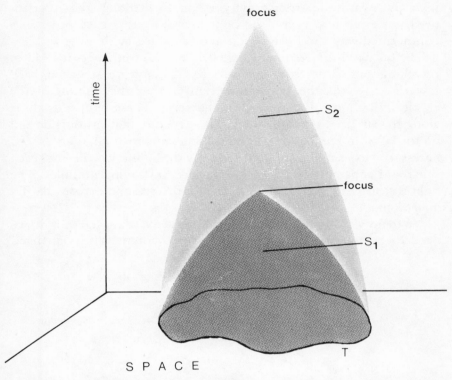

31 Trapped light. The wiggly ring T represents a closed two-dimensional surface in space that emits light. The inward-directed light falls to a focus along the cone-like surface labelled S$_1$, and the outward-directed light also falls inwards to a focus along S$_2$. These light surfaces therefore apparently completely enclose the region of spacetime between S$_1$ and S$_2$.

sucked inwards. To make the idea absolutely clear, Fig. 31 shows a spacetime diagram of this situation. Time runs vertically, and three-dimensional space is represented by two-dimensional horizontal slices. The wiggly ring marked T is our way of representing a closed surface surrounding the region of strong gravity. Light emitted inwards from this surface passes quickly towards the centre of the region, though it does not necessarily focus to an exact point because the system need not be spherical. These light beams are shown on the diagram as the short cone-like surface S$_1$. The shading on S$_1$ can be regarded as the paths of light pulses that leave points on T and plunge

73

inwards towards the strong gravity region. Even though the surface S_1 need not taper to a point, the light rays still cross each other in a complicated way and form a continuous surface, with no holes.

The outward-directed light rays behave very similarly. Instead of travelling outwards from the surface T, they are snatched back inwards, towards the gravitating centre, after the fashion of the elastically bound ball described in chapter 3. These light rays all lie along the surface S_2, another cone-like structure that sits outside and on top of S_1 in Fig. 31. The light once again is focused inwards, not necessarily to a single point, but certainly the light rays will cross each other and apparently close off the surface S_2, leaving no holes. The light is clearly trapped by gravity, and for this reason Penrose calls T a *trapped surface*.

Up to now, nothing has been said about a collapsing star, nor do we need to be too specific about this – the theorem is very general.

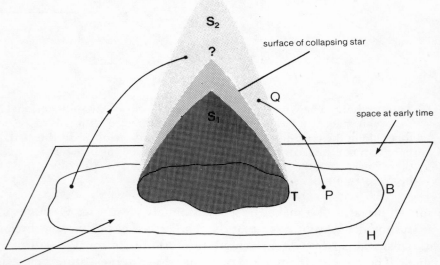

region of space that can influence collapse

32 Spacetime crisis. Everything that happens inside and on the surface of collapsing light, S_2, is determined by influences arising somewhere in the region of space inside the boundary B. But S_2 and S_1 together have no boundary (they form a closed surface). So where is the image of B on the light surface? Evidently something remarkable must happen to the topology of spacetime.

However, for the sake of clarity let us suppose that the focusing phenomenon shown in Fig. 31 is due to the collapse of a star, and draw the star in the diagram as well. This is shown in Fig. 32. The material of the star, being unable to travel faster than light, is compelled to lie between the two cones S_1 and S_2, and is therefore trapped: it cannot escape from this enclosed region.

At this stage we begin to see the importance of the simple observation that the light surfaces S_1 and S_2 apparently have no holes in them. Together these surfaces completely enclose the spacetime region in which the star is forced to shrink. In technical language, the light surfaces S_1 and S_2, taken together, seem to join up all around and hence to have the topology of a sphere (though the *shape* is nearer to that of two cones, fitting one inside the other). The star, being trapped inside this closed region, can never escape, even if it does *not* all shrink to a singular point of infinite density.

Penrose realized that there is something decidedly odd about the fact that S_1 and S_2 seem to have the topology of a sphere. (Of course, in our diagram, which has one dimension of space suppressed, the topology indeed looks like that of a sphere. In real spacetime, where there is an additional dimension, the light 'surfaces' S_1 and S_2 are really three-dimensional volumes. However, for ease of explanation we shall continue to talk of surfaces and to call the topology of S_1 and S_2 that of a spherical surface.) If attention is directed to the spacetime region at the bottom of Fig. 32, which represents a less traumatic epoch before the star has embarked upon its plunge to oblivion, then the whole universe at a single moment of time is to be represented by a horizontal slice H, extending infinitely in all directions, i.e. infinite space. According to naive ideas about causality, everything that happens after this early, quiescent epoch must be predetermined by what is happening at that moment, for all physical influences above the horizontal slice H either pass through H, or are ultimately caused by events on H. In particular, all the details of the subsequent gravitational collapse, including the curious focusing of the light rays along the cone-like surfaces, are completely determined by what happens on H.

The central feature of Penrose's theorem is the observation that although all events on S_1 and S_2 are caused by corresponding events on H, the topology of S_1 and S_2 is completely different from that of H, the former being the topology of a sphere, the latter the topology of a sheet, or infinite plane. This topological incompatibility reveals that something has gone very wrong in the spacetime region to the future of H.

It is worth examining this crucial but difficult point in detail. Events which occur on H a long way away from the collapsing star cannot influence what happens during the collapse itself because there is insufficient time for their signals to propagate to the collapse region, even at the speed of light. As no influences can exceed this speed, we must be able to draw a boundary (labelled B on the diagram) somewhere on H which encloses all of the events on H that can possibly exert an influence on the gravitational collapse of the star – even by rushing in at the speed of light. We can think of signals or influences leaving this enclosed region of H and falling in towards the star. In Fig. 32 the world lines of two of these influences have been drawn; they could be particles of matter, or photons, or any other kind of physical influence. For example, event P will send influences towards the collapse region and will determine, along with other events, what happens at Q on S_2.

It is easy to see at a glance that every point on the light surfaces S_1 and S_2 must be traceable back to some causative influence from the region of H enclosed by the boundary B. But this is absurd, because there is apparently *no boundary* anywhere on S_1 or S_2; the light surface has the unbounded topology of a sphere.

This conundrum recalls a more familiar problem of geometry concerned with geographical projections. The Earth has spherical topology, but a map is a sheet, like H. All maps have edges, i.e. boundaries, although there is obviously no boundary to the Earth's surface. The map-maker therefore has to cut the projection somewhere, usually through the Arctic and Antarctic. There is absolutely no way to map the unbroken surface of a sphere continuously on to an open sheet of paper.

Figure 33(i) & (ii) illustrates this point in detail. Suppose we have a translucent globe with the continents painted on the surface in some colour. The globe is suspended over a blank, flat sheet of paper on which the map is to be constructed. One geographical projection can be produced physically by literally projecting the images of the continents with a light on to the sheet of paper. Required for this purpose is a pinpoint source of light which, when placed near the

surface of the sphere, casts an image, or shadow, on to the surroundings.

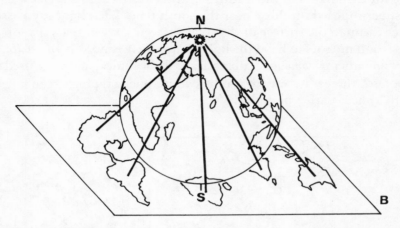

33 Map-maker's conundrum. There is no way to project all the globe
onto a sheet of paper with a boundary B, without overlapping
countries.
(i) When the light is situated inside the globe, some images near the
north pole are lost from the map.
(ii) When the light is outside the globe, the images are superimposed,
as each light ray passes twice through the surface of the globe.

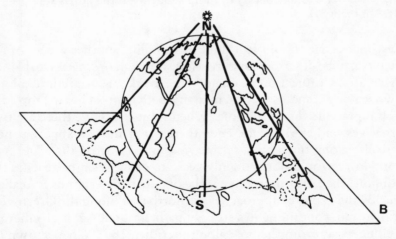

If the light is located inside the sphere, as in Figure 33(i), some of the
image will be projected upwards, and will not intersect the map at all.

To get the whole sphere projected down on to the map we must raise the light above it, as shown in Fig. 33 (ii). However, there is a snag in this, for the images of the features near the north pole will obviously be superimposed on those near the south pole. Each light ray passes twice through the surface of the sphere. Even if we employ elaborate lenses and mirrors to bend the light beam, there is no way of escaping this fundamental *topological* obstacle. In the technical language used in chapter 2, there is no way to produce a one–one mapping of the sphere on to a sheet with a boundary.

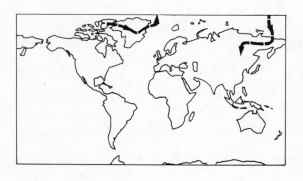

34 Two representations of an arctic itinerary.
  (i) Explorer's route appears to jump suddenly at the north pole.
  (ii) In reality the path is continuous.

Lest the reader should still doubt this point, another aspect of the difficulty can be described. Figure 34 (i) shows an ordinary world map in Mercator's projection. Consider the track of an arctic explorer shown leaving Canada. On the map this path approaches the top edge which represents the north pole. When it hits the edge there are two alternatives: either the track ends there, or it must reappear somewhere else along the boundary, moving inwards from the edge. As a real explorer would not suddenly cease to exist when he reaches the north pole, the latter alternative is the one that corresponds to reality. In Fig. 34 the explorer approaches the north pole along the Greenwich meridian and, continuing in a straight path, naturally enough emerges travelling away from the pole along meridian 180°. When shown on the surface of the sphere (Fig. 34 (ii)) the path is, of course, continuous. Only on the map with edges does it show the curious, artificial break.

The only way to render a map from a sphere to a bounded sheet completely faithful is to make a hole in the sphere somewhere. Suppose the Earth were hollow and there were really a hole at the north pole. The Earth would no longer have the topology of a sphere, because there is now a hole in it. In fact, a spherical surface with a single hole in it has the same topology as that of a sheet with boundary, as may easily be seen by envisaging the sphere to be made of a very flexible membrane. If the sides of the hole are drawn back, the sphere can be 'unfolded' and if spread out on a flat sheet will have, as its boundary, the edge of what started out as the little hole (see Fig. 35). If

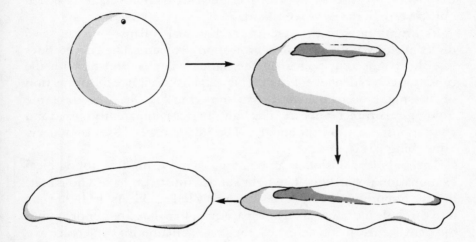

35 Unfolding a sphere. When the sphere contains a hole, however small, its topology changes to that of a sheet with boundary, as may be seen by opening it out. The edge of the hole becomes the boundary of the sheet.

the Earth were really like this there would be no problem about the tracks of explorers as represented on the map – this time the first of the two alternatives mentioned above would apply – the explorer's path would indeed stop when he reached the north pole, because he would fall down the hole.

Returning to Fig. 33 one can see how, when there is a hole at the north pole, a one-one projection can now be made on to the flat sheet. By locating the light precisely in the hole, all the light rays now pass only once through the spherical surface: no overlapping images are

79

produced, and the boundary of the hole becomes projected out to become the edges of the map.

It is important to realize that although we have been thinking of a small but finite hole, the topology of the sphere is altered as soon as just one single point is removed. A faithful, one-one, non-overlapping projection on to a map with boundary can be made with just one point missing – at the north pole, say. To fit the resulting image on to a small map, one would have to forsake the projection arrangement shown in Fig. 33 and employ some lenses to bend the light rays aiming for distant parts back on to the map. The image may be horribly distorted, but there is no problem in principle. The single point still gets opened out to become the edges of the map.

Returning now to the case of gravitational collapse, we can see a direct analogy with the Earth-projection problem. The curved lines like PQ connecting points on the light surface S_1 and S_2 with the causative antecedent events on H (Fig. 32) are rather like the projection of the sphere on to the sheet map shown in Fig. 33. And the same topological obstacle appears. The light surface is apparently closed and continuous – without boundary. The image on H has a boundary. Something has to give.

The only conclusion that one can draw from this topological examination is startling. The light surface must also have a hole in it, like the hypothetical hole at the north pole in Fig. 35. As the light runs into this hole it disappears from spacetime, like the explorer dropping out of existence. Just where the light goes is a matter for conjecture: we shall return to this point later. But clearly it cannot remain in the universe as we know it. Such is the extraordinary consequence of Penrose's theorem.

Penrose was not able to prove that the collapsing star always squeezes down to a point of infinitely dense matter. The Penrose 'singularity' is really a 'hole' or 'edge' of spacetime. An infinitely compacted star would certainly constitute a Penrose singularity, for when the density of the star becomes infinite the curvature of spacetime around it also becomes infinite. When that happens, spacetime is torn open and a hole appears. However, a hole in spacetime could conceivably occur even if the star does not become infinitely crushed. Therefore, when referring to singularities in what follows, we shall have in mind the wider definition used by Penrose.

At this point it is worth emphasizing the difference between a theory and a theorem, which is not always properly understood. When scientists formulate a theory about the world, it is essentially a conjecture – an educated guess, or a flash of intuition. An example of a long-standing theory is that matter is ultimately composed of many identical particles, or atoms. An alternative theory, now discredited, is that all material substance is continuous, no matter how small the scale on which it is examined. Often the word theory, particularly when used in common parlance, amounts to no more than that: a half-formulated idea. Police detectives, for example, often have a theory, without proof, about the identity of a criminal.

In scientific discourse, a proper theory should amount to much more than mere conjecture. To carry widespread conviction, the theory must be developed into a detailed predictive model, capable of being tested by experiment and sensitive to falsification. Physicists respect most those theories that have a rigorous mathematical framework which can be used to compute a wide range of effects. General relativity is such a theory. Einstein furnished a set of equations which anyone is challenged to solve in an attempt to check the theory's predictive ability. Either the equations are right, or they are wrong. The latter case can only be decided by experiment.

No amount of experiment, however, can prove the theory right, for it is always possible to conceive of an infinity of other theories that just happen to coincide with general relativity in those experiments that have been performed, but differ in others that have not yet been attempted. All such theories would be valid contenders for consideration. In this situation, the persuasiveness of a particular theory must always rest on other, essentially aesthetic criteria, such as simplicity and elegance. These additional features involve value judgements by the scientific community that cannot be rigorously defined or quantified. But it is in this restricted spirit that any description of the natural world becomes universally accepted.

Given that a theory, such as general relativity, is accepted on the basis of experimental confirmation, predictive power, vulnerability to falsification and general aesthetic appeal, the question then arises as to the logical consequences of the theory. If Einstein's equations are treated as correct, one may go on to examine the situations to which they inevitably lead.

This brings us to the topic of theorems. A theorem, as in the original Greek sense of the word, is a statement that is logically correct

81

provided a number of assumptions (or axioms) are first accepted without challenge. That is, given propositions A, B and C, it is possible to prove D. In this case, D is not a *theory*, open to falsification, but a *theorem*: a statement of fact. Naturally one may challenge the underlying assumptions A, B and C, but D cannot be challenged in isolation, for each step in the proof of the theorem should be a logical consequence of preceding accepted facts.

The classical application of logical deduction from accepted axioms is geometry. The axioms themselves seem above reproach (for example, through every distinct pair of points one may draw a straight line). A theorem, such as the celebrated one due to Pythagoras concerning right-angled triangles, is then a statement of fact, logically deduced from the axioms. It is, of course, possible, and even necessary when gravity is taken into account, to change one of the axioms of Greek geometry (as it happens, the axiom that through every point one may draw a line parallel to a given line is actually wrong), in which case Pythagoras' theorem may no longer be a correct statement of fact in the new system.

To return to Penrose's *theorem*, it should be thoroughly appreciated that Penrose is not *conjecturing* that spacetime has a hole in it. He has *proved*, on the basis of unassailable mathematical principles founded in universally accepted logic, that at least one light ray cannot continue indefinitely in a spacetime characterized by the existence of a trapped surface T; an initial space slice H and an assumption about the continuing convergence of the light rays. All three prerequisites may be challenged – and indeed we shall do so – but what is not at issue is the result, given these assumptions.

Before taking a closer look at the implications of Penrose's theorem, subsequent extensions of the work should be mentioned. The theorem was published in January 1965, and was followed by a flurry of interest among physicists. A whole series of papers then appeared in rapid succession, mostly by Penrose, Stephen Hawking and George Ellis of Cambridge University, and Robert Geroch at the University of Chicago. The purpose of these further developments was to widen the scope of the original theorem and, if possible, to weaken some of the assumptions without losing the essential result – that spacetime holes (not to be confused with black holes) are inevitable under a wide range of circumstances.

Much attention was directed to the subject of cosmology – the motion and behaviour of the whole universe – rather than the fate of a single star. Hawking was able to prove that trapped surfaces can occur around the entire cosmos. This could mean that all of creation is doomed to encounter a singularity. The implications of Hawking's result and other cosmological work will be discussed fully in chapter 8.

One of the results of the further work was the removal of the requirement that the space slice H in Fig. 32 be infinite in extent. Even in a finite universe where space is closed into a finite volume a singularity will still occur under very plausible circumstances.

What is one to make of these extraordinary mathematical results? It is possible to adopt a variety of positions. The first is to challenge the assumptions used in proving the theorems. As remarked, the one relating to the existence of an infinite space slice H at some initial moment is not in any case necessary. This leaves the question of whether a trapped surface will ever form in the real universe and, if it does, whether the convergence of the light rays along the cone-like surfaces S_1 and S_2 of Fig. 31 will continue all the way until the tip.

The formation of a trapped surface depends on the strength of gravity in a region of spacetime becoming so great that light cannot escape. In chapter 3 we saw that for a typical star this state of affairs would come about if the star were shrunk to the size of about a kilometre. Is this realistic? Stars are known which are only a little larger than this – the so-called neutron stars, which are thought to be the remnants of burnt out massive stars whose cores have collapsed under their own weight into a ball of neutrons. The further implosion of neutron stars is prevented by the existence of a subtle subatomic effect known as the Pauli exclusion principle. However, the Pauli principle can only support a certain total mass of neutrons, and if a star has a mass equivalent to, say, five suns, then it is known that further shrinkage must result.

Many stars are known with masses well in excess of our sun. Several stars having ten or more solar masses can be seen by glancing in the sky. In spite of this, one cannot conclude that at the end of its life cycle, such a star must retreat inside a trapped surface, i.e. terminate in a black hole. This is because various mechanisms exist whereby a star sheds material during its death throes. Frequently old stars blast off an outer layer into space, much as a snake sheds its skin: this may happen several times. Still more spectacular explosions occur in massive stars whose cores have burnt out and can no longer support their own weight. These co-called supernovae explosions will be described in detail in the next chapter.

Because the details of stellar gerontology are still obscure, no astronomer can assert with confidence that black holes will inevitably result at the end of a massive star's life. Nevertheless, on the evidence available it seems likely that a fair proportion of stars do end up inside trapped surfaces, and with stars being so numerous it is hard to believe that our own galaxy does not contain some of them. In the next chapter some of the current available evidence will be reviewed.

What concerns the physicist about the singularity theorems is not so much whether a large or small number of stars will actually end up producing singularities, but the fact that such a possibility is in principle possible. Even if it could be established that all normal stars do indeed escape the black hole fate, one can still envisage a technology powerful enough to create deliberately a black hole.

One straightforward, though demanding, technique for engineering a black hole is to exploit the fact that, as we shall see in the next chapter, the larger the mass of the stricken object, the lower the density of the material when it forms a trapped surface. For example, a star with the mass of our sun has to be compressed to above nuclear densities (a million billion grams per cubic centimetre) before a black hole forms. On the other hand, a star of one hundred solar masses will become a black hole at one ten-thousandth of this density. For a million solar mass object, a density of only one kilogram per cubic centimetre is necessary – extremely modest by astronomical standards. When one comes to a billion solar masses, the required density is below that of water. Thus, if enough stars were deliberately manoeuvred into a huddle, they could become a black hole without any individual stars even touching one another. Clearly, the possibility of a black hole is inescapable, whether or not they will actually occur in the real universe.

Assuming that trapped surfaces are to be taken seriously, then the only hope of avoiding a spacetime singularity would be if, once the light rays start to converge together in the fashion depicted in Fig. 31, something caused them to defocus again, so that they avoid intersection. This crucial focusing condition has been thoroughly investigated. The effect can be related to the gravity of material in the region of the cone-like surfaces S_1 and S_2. Crudely speaking, so long as the material here continues to attract with gravity, the light inexorably converges. The only chance of escape would be if *repulsive* gravity could ever occur.

Repulsive gravity is such an extraordinary idea that it is worth discussing in some detail. The delights of levitation are so deeply rooted in human society that it is clear that release from gravity represents an unusually compelling fantasy. Freud described the sensation of floating or flying as a recurrent dream image, and most people experience this dream from time to time. Legends of flying supermen, levitating Yogis and other religious devotees, as well as music hall 'floating lady' tricks, testify to a curious human interest in the idea of conquering gravity. Indeed, to this day some credulous folk can be persuaded to embark upon mysterious rituals and procedures designed specifically for the purpose of empowering them with ability to fall up rather than down.

But levitation is not only confined to tricks and myths. The science fiction writers H.G. Wells and Jules Verne both invoked levitating substances in an attempt to get otherwise weighty machinery aloft. Wells' 'cavorite' enabled the hero to visit the moon this way. Furthermore, somewhere in the back of most scientific minds (and somewhat nearer the front of military minds) is the faint hope that levitation – or antigravity as it is sometimes called – might one day become a reality. The technological implications of such a possibility are breathtaking. No more crude rockets or screaming aircraft engines, no more congested highways and dockyards – everything instead floating serenely from place to place.

Einstein's general theory of relativity makes quite precise predictions on the subject of antigravity, which unfortunately rules out completely all aspirations of levity along the above lines. The weight of a body is proportional to its mass. This is sometimes used to argue that if a body could exist with *negative* mass, it would have negative weight, i.e. would levitate, or fall upwards. However, this is a mistake. It is true that a negative mass would be repelled by the earth's gravity and experience an upward force, but because its mass is negative, such a body would also have negative inertia. That is, when the body is pushed in a certain direction, it will move in the opposite direction. Clearly, an object like this could not be composed of ordinary substance, for if one shoved it with a hand, it would interpenetrate one's arm. When released above the ground, a negative mass will, in response to the upward force of levitation, fall downwards in the usual way. So much for cavorite!

The existence of negative mass would have other curious effects. If two bodies are placed side by side, one being an ordinary body with positive mass, the other a body with an equal quantity of negative mass, then their subsequent behaviour is most unusual. A force of gravity acts between the two bodies, but because one has negative mass the force is in fact repulsive, i.e. levitational. As explained above, the negative mass, with its negative inertia, will fall towards its companion in response to the levitational force. The ordinary (positive) mass, on the other hand, will also feel a repulsion, but enjoying ordinary (positive) inertial properties, it will move away from its companion under the reaction of the levitating force. Thus, both masses move in the same direction. As fast as the negative mass drops towards the positive mass, so the positive mass slides out of its way, and the remarkable pair chase off across the universe at ever-increasing speed and fixed separation. Now *that* is a good engineering possibility!

Although when dropped in the gravitational field of a much larger object, a negative mass will not behave unusually, nevertheless the gravitational force which it exerts on other positive masses is repulsive, as explained above. If the Earth had a negative mass, then we would indeed fall upwards in its vicinity.

To return to the subject of gravitational collapse and the singularity theorems, a crude way of describing how the light rays might be bent back out again before hitting the singularity is if somehow the collapsing star acquired a negative mass as it approached a very high density. The onset of this levitating force would repel the falling light rays outwards, and spacetime would be saved. How seriously can the idea of negative mass be taken?

First it is necessary to note that according to the general theory of relativity mass is not the only source of gravity. Pressure, stress and energy also contribute. If any of these quantities become sufficiently negative, then levity will occur. In recent years a certain amount of research has been undertaken by physicists attempting to discover situations in which negative energy might arise. It has long been known that certain subtle subatomic processes can in principle produce negative energy, and the question is whether or not these processes would occur in the extreme conditions associated with gravitational collapse. As we have absolutely no hope of ever experimenting with a collapsing star, the question must remain open, with only mathematical theory as a guide. At the time of writing the theoretical work predicts some circumstances in which singularities would be avoided by negative energy or pressure, but the circumstances are rather artificial.

If a singularity does not form, or if some of the collapsing matter misses the singularity, a mystery surrounds what becomes of it. As explained in the previous chapter, being inside an event horizon, the matter cannot escape back into the universe we know, yet if it does not disappear at a singularity it must pass into a region of spacetime that we do not know. Sometimes such unknown regions are called other universes, the effect of gravitational collapse being to establish a bridge or tunnel into these enigmatic 'parallel' worlds. The collapsing star, or whatever matter falls in after it, falls on through the tunnel and out into, presumably, a cosmos much like our own.

It is not really known what the other end of the tunnel would look like. Sometimes it is conjectured that the matter which falls inwards in our universe gushes outwards in the other. Conversely, the collapse of a star or other object in the other universe would appear to us as an explosive eruption. There is no lack of exploding astronomical objects that could fit this speculation, an especially good class of candidates being the so-called quasars (acronym for quasi-stellar radio source) which will be further discussed in the next chapter.

There is no reason to suppose that only one other universe is involved; each collapsing star could connect us with a different cosmos. Indeed, in one idealized model of an electrically charged or a rotating black hole, the singularity does not block off the whole spacetime – crudely speaking there is a hole or tunnel through the middle of it – and an observer could fall right through it into another universe. However, if he chose to fall back in again, he would not re-emerge from the black hole in our universe, for that would violate causality. Instead he must travel on to a third universe. Repeating the infall, he would reach a fourth, then a fifth and so on. It seems that an infinity of other universes are required if this model is taken seriously.

Many physicists feel distinctly uneasy about the idea of other universes joined on to our own through the interiors of black holes. It is often considered preferable to suppose that all matter that falls inside a trapped surface will necessarily end up hitting a singularity and disappearing from spacetime, although the singularity theorems themselves make no prediction on this matter.

Precisely what, one is prompted to wonder, becomes of the matter that hits a singularity? The question is really philosophical rather than physical. The sorts of singularities about which physicists have most understanding are those that involve the curvature of spacetime growing without limit, such as would occur if an exactly spherical star collapses. Infinite curvature implies infinite gravitational forces, so the approaching matter would suffer unlimited violence. First, the matter nearest the singularity would be attracted more ferociously than that farther away, so the material would be ripped apart. Secondly, the inward rush would squeeze the material into an ever diminishing volume of space until it became shrunk to nothing. So the sacrificed object would be simultaneously ripped apart and crushed. All structure, however robust, would eventually succumb to the immensity of the singularity's gravity.

In spite of the fact that the falling object is utterly destroyed, it is hard to resist the impression that something – some shattered, compressed debris that may not even be called matter – must survive. Such, however, appears not to be the case, for taken at face value the singularity marks the end of the physical world and the end of physics, as well as the end of space and time. In short, when matter runs up against a true spacetime singularity, it has reached the edge of existence itself. There can be nothing 'beyond' that is in any sense relevant to the affairs of our world, for no influence whatever can cross the barrier of the singularity. Of course, one can imagine all manner of things to be located on the far side of a singularity, but that is idle speculation, because no observation or experiment can ever be performed to test the idea, or to support one picture rather than another. The singularity seems to mark the edge of anything that we can know, and anything that is meaningful in our world.

Not all physicists, however, accept the existence of true singularities, even as a remote possibility. They argue that when the curvature of spacetime reaches a sufficiently high value, the general theory of relativity, upon which the singularity theorems are based, is no longer an accurate description of reality. Indeed, the very notion of spacetime may cease to have meaning under these excessively violent circumstances.

For example, the continuity of space and time has been questioned by some. In chapter 2 it was explained how a line is regarded by mathematicians as infinitely divisible, containing more points than all the infinity of whole numbers. No physical experiment, however, could ever verify that the mathematician's idealization of a continuous line corresponds to the topological properties of real space. Suppose, for example, that space really were made out of lots of little 'lumps' joined together – atoms of space with no internal structure, strung together in a lattice. A similar concept could be invoked for time: each event would be connected to the next in a jumpy and discontinuous fashion, like the frames of a movie film. However accurate our perceptions may become, however fine the resolution of our scientific instruments may be, there always exists a smaller scale of space and time inside which this sort of discrete latticework could exist. Nobody has the slightest idea how a collapsing star would behave in such a latticework. Would the star end up wandering among the points of the lattice, *inside* spacetime?

The notion of 'atoms' of time has a long history. In the twelfth century the Jewish philosopher Moses Maimonides wrote: 'Time is composed of time-atoms . . . which in fact are indivisible.' It is possible that he was drawing on older sources from India in this assertion. The Franciscan monk Bartholomew the Englishman was quite explicit when he wrote circa AD 1230 that there are 22,560 time-atoms in an hour. As modern clocks can happily tick away over intervals far less than a millionth of a second, this courageously low figure can be discounted. In later years no less a thinker than Descartes also subscribed to the time-atom hypothesis, contending that ordinary matter would remain trapped at each instant, lacking the capacity for endurance, were it not for the continual intervention of God.

Many physicists believe that it is wrong to regard space and time as primitive entities anyway. For example, John Wheeler regards spacetime as rather like a block of elastic material. On a large scale elastic appears continuous and deformable, but on closer exmination one finds that it is really built out of atoms, and the deformation is due to the changing arrangement of the atoms relative to each other. Wheeler conjectures that there may exist a 'pregeometry' out of which space and time are built in the same way that the elastic medium is built up out of atoms. Of course, this need not necessarily imply that singularities are avoided, for one can experience a piece of elastic snapping if it is twisted or torn violently.

It is generally accepted that some modification of the usual notions of space and time must inevitably occur when ultra-small distances and intervals are considered. This belief stems from the acceptance of a thoroughly tested and extremely successful theory of microscopic systems called the quantum theory, mentioned briefly at the end of chapter 2 and described fully in my book *Other Worlds*. The basis of this theory is that in nature there is an inherent uncertainty or unpredictability that manifests itself only on an atomic scale. For example, the position of a subatomic particle such as an electron may not be a well-defined concept at all; it should be envisaged as jiggling around in a random sort of way. Energy, too, becomes a slightly nebulous concept, subject to capricious and unpredictable changes.

One effect of these microscopic fluctuations is to prevent the collapse of atoms under the action of electric forces between the nucleus and surrounding electrons. It has been conjectured that perhaps gravitational collapse may also be prevented when a star has shrunk down to submicroscopic dimensions, and quantum theory starts to play a role. It is easy to visualize how this may happen. The size of the quantum energy uncertainty increases as the scale of distances decreases. Across an atom, energy is likely to fluctuate by only a few electron-volts (an electron-volt is the energy acquired by a single electron accelerated by a one volt electric potential). From the gravitational standpoint that sort of energy is trifling. Nevertheless, over ultra-small dimensions – about a million-billion-billion-billionth of a centimetre – the energy uncertainty is so large that its gravitational effect dominates over that of any matter. So strong is the quantum-induced gravity that it may even bend spacetime into a froth of wormholes and bridges; rather than a continuum, spacetime would resemble more a sponge-like structure with a highly complicated topology. The unit of time inside which these bizarre structures are important is almost inconceivably small – a ten-million-billion-billion-billion-billionth of a second.

If these ideas are even remotely right, it is clear that the details of gravitational collapse and spacetime singularities will be profoundly modified when the star shrinks to these unimaginably minute dimensions. In that case, a singularity need not necessarily be regarded as an edge to the physical universe but, less drastically perhaps, as an edge of space and/or time only. What lies beyond space and time we do not yet know. Maybe it is a lattice, or a spongy foam, or some other completely alien and abstract structure yet to be discovered.

Laboratory experiments have so far only tested the traditional, continuous idea of space and time down to dimensions about one thousandth of an atomic nucleus – that is, about a billion-billionth of a metre, which is still a billion billion times larger than the scale at which quantum effects should appear. It is entirely conceivable that with the ever-higher energies available in 'atom smashers', subatomic probes could discover 'atoms' of space and time in the near future, over lengths well in excess of the 'wormhole' region. Whether or not this is the case, the spacetime singularity will always be regarded by some as simply an indication that it is familiar notions, rather than the physical world, that come to an end there. Either way, the singularity is certainly nature's greatest crisis.

5 Black holes and the cosmic censor

Once the mathematical theorems about the formation of spacetime singularities had become known, physicists began to worry about the implications. The appearance of a singularity in spacetime will, as we shall see in the next chapter, have dire consequences for the physical world. How can the universe be made safe from such a monstrosity?

In the original Penrose theorem, extensive use was made of the properties associated with the existence of the so-called trapped surface discussed in the previous chapter. As the name implies, the trapped surface prevents any light or matter leaving the spacetime region near it and travelling out to some distant observer. Consequently, without falling into the collapsing system himself, this observer can never witness the late stages of the collapse or the ultimate annihilation of spacetime and matter at the singularity. The information about this violence, which must be conveyed to the observer in the form of signals carried by material particles, light, or some other radiation, is also trapped. As mentioned in chapter 3, the inability of light to escape from some region of the universe means that a distant observer regards that region as a black hole. If all singularities occur only inside black holes, the rest of the universe will be spared any of the unpleasant consequences.

Although Penrose proved that a trapped surface was a sufficient condition to produce a singularity, he did not show it was necessary. The possibility remains that a singularity will form even in the absence of a black hole. Such singularities have become known as naked, and a great deal of investigation has been undertaken to determine whether singularities can ever occur naked, or whether they will always be safely hidden inside black holes.

In 1969 Penrose formulated what he called a 'cosmic censor hypothesis': 'We are thus presented with what is perhaps the most fundamental unanswered question of general-relativistic collapse theory, namely: does there exist a "cosmic censor" who forbids the appearance of naked singularities, clothing each one in an absolute event horizon?' In picturesque language this idea may be expressed as follows. Nature abhors singularities, and always hides them from our eyes by clothing them in a black hole. The full nakedness, with all its shocking implications, can never be revealed to us.

Does a 'cosmic censor' really exist to prevent naked singularities? Has nature really got a built-in escape mechanism to hide the singularity from the rest of creation? Much effort has been devoted to searching for such a mechanism or, alternatively, trying to prove that there is none by constructing a mathematical model in which a naked singularity is definitely predicted. This work continues today, with Penrose's fundamental question still unanswered. As we shall see in the forthcoming chapters, the consquences of an uncensored universe are likely to be bizarre indeed.

In the absence of a direct observation, two conditions must be met before the idea of naked singularities can be taken seriously. The first is the existence of some explicit examples where calculations performed on mathematical models using Einstein's general theory of relativity lead to the formation of a singularity without the accompanying event horizon. The second is some definite astronomical evidence that the sort of catastrophic gravitational collapse that is being discussed in this book is likely to happen in the real world. In this chapter, the astronomical evidence will be reviewed; the next chapter will deal with some of the theoretical ideas about the formation of naked singularities.

Because physicists and astronomers are fundamentally sceptical, as befits scientists faced by a bewildering array of inexplicable phenomena, they are inclined to be wary about the very idea of a body suddenly imploding to nothing under the action of its own gravity. A naked singularity, being one step worse than a black hole, is considered the greater of two evils, so that current research directed at detecting gravitationally collapsed systems is almost entirely concerned with locating and observing the effects of black holes. Hence, the currently available evidence for black holes is the topic to which we shall now turn. If, as is the case, the evidence suggests that black holes are real, then we know that the possibility of naked singularities must also be taken seriously, for it seems likely that, in the absence of a mysterious cosmic censor, the sort of situation that leads to the formation of a black hole could, with a little disturbance, give a naked singularity instead.

The earliest speculation about what are now called black holes dates from the late eighteenth century when the French mathematician and astronomer Pierre Laplace reasoned that there could exist stars that are so massive and/or so compact that light will not be able to escape the grasp of their gravitational fields. These black stars were not necessarily supposed to implode, as we now know must be the case on the basis of the theory of relativity, but the fact that Laplace guessed the existence of black, intensely gravitating objects nearly two hundred years ago is surely a remarkable piece of scientific prediction.

Born in 1749, Pierre Simon Marquis de Laplace was the son of a farmer and became one of the leading European minds of the late eighteenth century. As professor at the École Militaire in Paris, he administered Napoleon's entrance examination in 1785 and became adept at ingratiating himself with the political establishment. He thereby gained a great reputation in French society, receiving a title and much wealth. His sparkling intellect ranged across many fundamental problems in the physical sciences, but he is probably best known for his work on celestial mechanics, especially his 'nebula hypothesis' for the formation of the solar system which became widely used by astronomers. Being something of a philosopher, he is recorded as having told Napoleon 'I have no need of this hypothesis' when discussing the existence of God. Indeed, in 1773 he presented a paper to the Academy of Sciences in which he proved mathematically that the solar system was stable. Newton, who held some strong and mostly unconventional religious views, had proclaimed that divine intervention was sometimes necessary.

On the subject of black stars, Laplace's reasoning is simple, elegant and easy to follow. People often say 'what goes up must come down' but in these days of space travel we know, as Newton did even in the seventeenth century, that it is not true. Because gravity weakens its grip farther from Earth according to the inverse square law discussed in chapter 1, it follows that if a body can be projected fast enough from the Earth's surface it will rise beyond the strong-gravity region and thereafter only suffer minor, and diminishing restraint. It will then never return. The minimum velocity required for a body to escape the Earth's gravitational field and disappear irretrievably into space is known, appropriately enough, as the escape velocity. It works out at about 11½ km per second at sea level. This is the speed required of a rocket that is, for example, to travel to the moon or beyond.

All astronomical bodies have an escape velocity, some greater than the Earth's, some less. The magnitude is determined by two parameters: the total mass of the body, which determines its gravitating power, and the radius of the body, which determines the strength of the gravitational force at its surface. A high escape velocity is produced by a massive and/or compact body. For example, the escape velocity from the surface of the moon is only 2.4 km per second, while from Jupiter it is 59.6. From the sun it is no less than 618.2.

What Laplace realized was that if a star is massive enough, this number might rise as high as the speed of light (about 300,000 km per second). Under those circumstances, he argued, 'the attractive force of a heavenly body [would] be so large, that light could not flow out of it'. If the escape velocity exceeds the velocity of light, then light cannot escape. When could such a state of affairs be expected to come about?

In the previous chapter it was remarked that if millions of stars were deliberately manipulated into a compact group, they would collapse together into a black hole without the need for further compression. A glance at an astronomy encyclopaedia reveals many examples of large clusters of stars in high concentrations. For example, our galaxy contains dozens of so-called globular clusters, containing up to a million stars each, grouped together in a dense ball. At the centre of a globular cluster the average spacing between stars is only about one tenth of a light year, compared to several light years in our neighbourhood of the galaxy.

The motions of the stars in a globular cluster are complicated because of all the mutual gravitational forces acting between them. However, one general feature can be deduced. Although actual physical collisions between individual stars are rare, fairly close encounters will frequently deflect stars from their paths. The stars in the core mill around more or less at random, unable to escape into galactic space by the gravitational pull of the rest of the cluster. Occasionally, however, due to the fortuitous alignment of motions and the accidentally advantageous arrangement of collisions, some stars will be imparted

95

with somewhat more energy than others and hence move faster. A few will acquire enough speed to quit the cluster altogether and escape. Slowly but surely, over hundreds of millions of years, stars 'evaporate off' from the central region.

As with a liquid, evaporation causes the residue to cool in order to supply the extra energy to the fleeing particles. Cooling a star cluster means slowing up the stars, so that they are unable to sustain enough collective agitation to resist the ever-present pull towards their centre of gravity. Thus, to supply the energy of the ejected stars, the core must shrink, and this brings the stars in the crowded centre of the globular cluster still closer together. This suggests that such clusters only have a finite lifetime, at the end of which they collapse into a black hole, perhaps surrounded by a diffuse halo of stars. Calculations suggest that it takes between ten million and ten billion years for the collapse to occur, and as observations put the ages of globular clusters at ten billion years (i.e. about the age of the galaxy – see chapter 8) one may conclude that many of our galaxy's globular clusters have long ago collapsed to black holes. The clusters that we see now are only a long-lived vestige that still survives. It is to be expected that hundreds of black holes with masses of, say, one thousand suns, lurk inconspicuously around the galaxy – the remnants of ill-fated globular clusters.

Not knowing what a dead globular cluster looks like, astronomers have directed their attention to searching among live ones that might still be in the process of collapsing into black holes. The idea is that once a small 'seed' hole forms at the centre, it will draw other stars into it irresistibly, so that the distribution of stars in the core of a globular cluster containing a black hole will display an otherwise unaccountable concentration of stars close to the centre where the hole is located. So far, telescopic searches have failed to yield decisive evidence for such an arrangement.

Another place where stars are strongly clustered is in the centres of galaxies, and it has long been speculated by astronomers that massive black holes form there. Unfortunately, the centre of our own galaxy is badly obscured by gas and dust, and it would be extremely difficult to detect by traditional means the effects of black holes that, in spite of their mass, are likely to be smaller than the solar system.

One very recent technological development could, nevertheless, be used to 'see' through all the obscuring material right into the heart of the galaxy. This is the so-called gravity wave telescope. Most telescopes operate by detecting and analysing electromagnetic waves, such as light, radio waves or X rays. A gravity wave telescope responds to waves in the gravitational, rather than electromagnetic, field. Gravity waves are produced whenever large quantities of matter are violently disturbed in some way, such as when a star falls into a massive black hole. It is possible to envisage an energy equivalent to ten times the total output from a star over its entire multi-billion-year lifetime being emitted by the stricken object in a pulse of just one'tenth of a second duration. The gravitational 'luminosity' of such a source is around a billion billion billion billion billion kilowatts.

In spite of the colossal release of energy expected from the violent encounter of a black hole with stellar debris at the galactic centre, gravity waves interact so feebly with matter that they are almost impossible to detect. In the next chapter, some of the experiments that are currently in progress to achieve this goal will be described.

One other possibility is to try to spot a massive black hole at the centre of another galaxy, rather than our own. Paradoxically, it is easier for us to detect its effects from a great distance than it is when the hole is immersed in the complexities of our own galaxy. In 1978, astronomers from Britain and the US made a careful photometric study of two galaxies known as 'ellipticals' because of their characteristic squashed-ball shape, contrasting the more usual spiral structure. One of these, called M87, had been spotted and catalogued two hundred years ago by the French astronomer Charles Messier (hence the label M). It had been known for some time that there was something a bit odd going on in the centre of M87, and the new observations confirmed this. The astronomers compared the organization of stars in M87 with those in another elliptical called NGC3379, and found a dramatic difference. Naturally enough, in both cases the density of stars was found to increase steadily towards the centre of the galaxy, for there is a tendency for stars to congregate near the centre of gravity. What was remarkable, though, was the way in which the positions and motions of the stars near the centre of M87 appeared to be severely disturbed by an unseen and compact massive object, precisely as one would expect if a huge black hole were exerting its gravitational influence on the stars in its vicinity. Such dissimilar density profiles in two otherwise similar galaxies led the astronomers to suggest the strong likelihood of an enormous black hole in M87, with a mass of about five billion suns.

Just before their paper was published in *The Astrophysical Journal* the news of the result was leaked to the press and it was widely hailed as the definitive discovery of a black hole. While such a pronouncement is probably premature, it does seem quite likely that M87 really does contain one of these enigmatic objects. No doubt future observations will decide the issue.

Looking still farther afield, the mystery objects known as quasars (acronym for quasi–stellar objects, because they look at first glance like nearby stars) have long been the subject of black hole speculation. Indeed, it was after the discovery of quasars in the early 1960s that astronomers began to take the idea of black holes seriously – though the phrase was not coined until 1968.

Quasars, which seem to be associated with the elliptical galaxies, are believed to be among the most distant objects in the visible universe, which makes them hard to study as they are also very small. Estimates put the size of the active central region at little more than a few times the size of the solar system. Yet erupting out of this tiny volume is a flow of energy exceeding that of a whole galaxy. To supply such a colossal output of energy requires a total mass of at least one hundred million suns. The concentration of this quantity of matter into a volume as small as a quasar brings the system very close to the brink of gravitational collapse.

Although there are several rival models that do not involve black holes at all, it would be surprising if some quasars did not contain them. Astrophysicists envisage a halo or disc of gases surrounding the hole, steadily falling inwards, and perhaps spiralling around. Complicated viscous, thermal, acoustic and perhaps magnetic processes then generate and transport vast amounts of energy outwards through this infalling material, from where it pours into space in the form of radio waves, light and other types of radiation. If the central hole is rapidly rotating the outbursts of energy might occur along a preferred direction. Many quasars and exploding galaxies are known which have jets or eruptions in particular directions.

The search for black holes in the centres of globular clusters, galaxies and quasars is motivated by the knowledge that an event horizon will form around a large enough group of stars without the stars themselves even touching one another. This is because the density of material needed to produce a black hole diminishes with the total quantity of matter. It therefore seems inevitable that, with the natural tendency of star clusters to shrink, black holes will be the end product.

A second scenario for the formation of black holes is quite different and involves individual stars rather than whole clusters. In this case, the density of matter must become enormously large if an event horizon is to form. In the case of the sun, it has already been remarked that the entire mass of the sun must be shrunken into a ball only about three kilometres across before a black hole will form. Can such an extraordinary degree of compaction ever occur to a star? When the idea was first proposed in 1939 by J. Robert Oppenheimer and H. Snyder, it was not treated seriously by astronomers. Indeed, it was only in the late 1960s, with the accidental discovery of the so-called pulsars, that the possibility of the collapse of single stars into black holes left the realm of science fiction.

Pulsars are observed as rapidly pulsating radio sources, flashing on and off perhaps dozens of times a second, with extreme regularity. There is no known way that an ordinary star, or even a white dwarf star about the size of the Earth, could pulsate so rapidly. The highly regular pulses suggested the presence of an object that was rotating, emitting a burst of radio waves on each revolution. Very soon astronomers concluded that the radio emissions from the pulsar were in the form of a beam. Each time the object spins round, the beam sweeps past the Earth, like a lighthouse, giving a brief radio flash.

To rotate up to 30 times a second required an object only a few kilometres across, otherwise the enormous centrifugal force would fling material off the equator and the 'star' would come to pieces like an exploding flywheel. It was proposed that the central object that drives each pulsar is the remnant of a star that has imploded under its own gravity and become so compact that even its atoms are crushed. Instead of consisting of ordinary matter, this collapsed star is more like a grotesque atomic nucleus, consisting almost entirely of neutrons – a neutron star. So compressed is the material that a thimbleful of neutronic matter would contain the equivalent of a billion tonnes of Earth matter (say, a cubic kilometre of rock). More bizarre still, the surface gravity on such a shrunken object would be many billions of times greater than on Earth, so that a thimbleful would weigh more than all the continents of Earth together!

Once neutron stars were taken seriously, black holes were not far behind. Further compression by only a small factor would convert a neutron star into a black hole. Theoretical studies of neutronic material suggested that if the total mass of the neutron star were only slightly greater than the mass of the sun, then the rigidity of the neutronic material would be unable to sustain the huge weight of the shrunken star in the enormous gravitational field: catastrophic collapse would ensue.

Neutron stars and black holes came to be treated as the natural end state of a star's life cycle. The existence of any star can be regarded as a struggle between competing forces. On the one hand gravity tries to pull all the material down into a smaller and smaller volume, while on the other hand the internal pressure due to the extreme central temperature tries to blow the star to pieces. For most of its life these competing forces manage to balance each other and peacefully co-exist. When a star gets old, however, instabilities set in and the star may either explode or implode (or both) as the forces become unbalanced.

In a middle-aged star like the sun, the internal heat that supports it against gravity is generated by nuclear processes in the central region, or core. Astronomers were long mystified as to the source of the solar heat. Spectroscopic studies reveal that most of the sun is made of hydrogen, but even with such a highly reactive gas, ordinary burning could not keep the sun alight for more than a few million years. As the Earth contains fossil remains of life dated at billions of years, there clearly exists a source of energy in the sun more powerful than can be supplied by ordinary chemical processes. With the discovery of nuclear power, the answer was found. The sun, and most stars, are huge nuclear furnaces, operating by fusion power, as in the case of the hydrogen bomb.

In ordinary burning, whole atoms rearrange themselves into new types of molecules, releasing energy as a consequence. For example, when wood or coal burns in air, the carbon atoms contained in this fuel fuse with oxygen atoms in the air to form molecules of carbon dioxide (one carbon atom stuck to two oxygen atoms). The force that sticks them together is electromagnetic in origin, and derives from the fact that atoms contain electrically charged particles (electrons and protons). When nuclear burning takes place, it is the central nuclei of atoms that rearrange themselves. In the sun, for example, hydrogen nuclei (protons) fuse together into the nuclei of the element helium.

The force that glues the nuclear particles together is hundreds of times stronger than the electromagnetic forces that bind atoms into molecules, and the energy released is correspondingly greater. By converting hydrogen into helium, the sun can keep up its present heat output for ten billion years. And there is no shortage of hydrogen fuel in the universe. About 75 per cent of the cosmos is made of it.

Even when a star has exhausted its hydrogen fuel its life is not over. The further fusion of helium nuclei into carbon and other elements can continue to supply a great deal of energy for a long time. However, to extract this additional energy, the central temperature of the star must go higher, for the heavier fuels are harder to burn than hydrogen. This will happen automatically, because when the hydrogen in the star's core is exhausted, the fusion reaction stops, the pressure begins to fall and gravity starts to win the ever-present struggle between the forces of compression and expansion. The core shrinks and, as with all compressed gases, the temperature rises. Shrinkage continues until the temperature climbs high enough to ignite the helium, and away the star goes into the next cycle of consumption.

One by one heavier nuclei are built up by the fusion of lighter ones, until a point is reached, round about the element iron, where the energy gain from further nuclear burning turns into a deficit, and the star reaches the end of the road. No further nuclear fuel is available to sustain the internal pressure needed to support the star against its own weight. Gravity, always waiting in the wings to crush unsupported matter, grasps the faltering core and implodes it almost instantaneously. Though this is a very simplified picture of events in the centre of the star, the upshot of the core's implosion is to blast away the outer layers of the star with a tremendous release of energy, leaving the core as a shrunken cinder, either a neutron star or a black hole. Thus, the almost unimaginable compression needed to force a star to become a black hole can indeed be found in the violence that accompanies the death of a star.

Not all stars will die by simultaneous explosion–implosion, but it is believed that the more massive ones do. The outburst that accompanies the suicide is conspicuous indeed. For a brief duration, a single star may outshine a whole galaxy. Known as a supernova, an event like this was recorded by oriental astronomers in 1054. The shattered debris of the star that erupted is still visible as a ragged cloud of exploding gas, known as the Crab nebula, located in the constellation of Taurus. As expected, amid the nebulous gases is a pulsar, the neutron star being all that remains of the original interior of the star.

Estimates put the number of supernova events per galaxy at around three per century, though the last to be recorded in our own galaxy was as long ago as 1604. It is not known how many such events produce black holes rather than neutron stars, but it would be surprising if it were not a fair proportion. It therefore seems reasonable that, over the ten billion years that our galaxy has existed, millions of black holes have formed from the death throes of old stars.

The heavier a star is, the more likely it is to form a black hole, for it is known from the theory of relativity that a collapsed object with more than three solar masses cannot avoid becoming a black hole. When the star explodes, it may shed a sizeable fraction of its total bulk, so the greatest chance for black holes rests with stars that start out with ten or more solar masses. Many such stars are known. Moreover, the heavier stars are also the ones with the shortest lives. The extra weight causes greater compression in the middle and hence higher temperatures and pressures to support it. The higher temperature burns up the nuclear fuel faster. A star of ten solar masses might only live for a hundred million years. Consequently, the galaxy should be replete with the remnants of such objects, many of which began their life cycles as long as ten billion years ago.

All these considerations have led astronomers to believe that black holes are rather common objects in the galaxy, particularly near the galactic centre or in globular clusters where there is a high concentration of very old stars. But how are they to be detected?

As a black hole cannot be seen directly, its detection depends upon spotting secondary effects. Having the mass of a star, it will produce intense gravitational disturbance among its neighbours. This is especially true of a black hole in a so-called binary system. A great many stars in our galaxy are not wandering about individually, but are grouped in twos or even threes. Two stars can remain close together, without falling into one another, by orbiting around their common centre of gravity. Binary star systems are known with orbital periods varying from hours to years, depending on the separation between the components.

If one member of a binary star pair dies and becomes a black hole, the other will continue to orbit about it in the usual way, but from

Earth it will seem as though the companion star is revolving around empty space. The black hole is too small to be seen. Systems of this type have long been known to astronomers, but some confirmation has to be found to ensure that the unseen companion is not simply an inconspicuous ordinary star, or a neutron star.

In the 1970s, with the growth of satellite astronomy, such a means of confirmation came to hand. As the black hole revolves around its companion star, it will raise tidal bulges due to its intense gravity. If the hole gets too close, it will succeed in actually ripping material out of the surface of the companion. This debris will fall towards the hole, but the orbital rotation of the system will encourage a vortex of gases to establish itself around the hole. The material near the centre of the vortex is steadily swallowed by the hole itself, like water disappearing down a plug hole. As it slowly spirals inward, the gas will become very hot; so hot, in fact, that it will not only glow, but will emit energetic X rays. In the early 1970s astronomers began to search for X rays from binary stars with unseen companions, using satellites containing X ray telescopes.

Shortly, one very good candidate emerged. Known as Cygnus X-1, this is a system located in our galaxy in the constellation of the Swan. Through a conventional optical telescope astronomers can see a single star known as a blue supergiant on account of its blue colour (indicating it is considerably hotter than the sun) and its great size (much larger than the sun).

By studying the spectrum of the star's light it is possible to infer that it is in rotation about an unseen companion. Such a conclusion is reached because the supergiant, as it orbits around the invisible star, sometimes moves towards us and sometimes away from us at about 300 kilometres a second. When advancing, the light-waves from its surface are bunched up a little. This phenomenon, known as the Doppler effect, was mentioned at the beginning of chapter 3. The shortened wavelength produces a higher frequency. In the case of light waves the high frequency is manifested as a slight change in the colour quality of the light (a shift towards the blue end of the colour spectrum). Conversely, a receding star is slightly reddened as its light frequency falls. Thus, by looking for a tiny wiggle in the colour of the blue supergiant, it was established that the star orbits its invisible neighbour once every 5½ days. In addition to the primary wiggle, there are secondary disturbances that have been interpreted as due to gas being dragged off the supergiant on to the companion object in the fashion explained above.

From an orbiting X ray telescope, Cygnus X-1 appears as a single bright source that fluctuates somewhat at long wavelengths. It seems that from time to time the supergiant gets partially in the way of the X ray source and eclipses it, which suggests that the source of X rays is close to the unseen companion. Although the source does not pass right behind the supergiant from our particular line of sight, some of the gas streaming off the large star is bound to obscure the X ray source on occasions.

All these observations make it appear very likely that the dark companion star in Cygnus X-1 is either a black hole or a neutron star. The only real hope of distinguishing between them is by weighing it; neutron stars cannot be heavier than a very few solar masses.

How does one weigh a star that is many light years away? Several methods are possible, the most direct being an appeal to the laws of motion for orbiting bodies, originally discovered by Kepler and explained by Newton. The basic data needed to perform the calculation are firstly the orbital velocity of the supergiant, secondly the size of the supergiant and thirdly some information about the inclination of the orbits relative to our line of sight.

The first piece of information is provided by a study of the spectrum of the supergiant: the faster it moves, the greater the colour disturbance. The radius of the star can be deduced from a knowledge of its temperature and its luminosity, because one can calculate how brightly a given area of material should glow at any given temperature. Knowing how brightly the whole star radiates enables one to compute its surface area, and thus to calculate the star's radius. The temperature of the star can be inferred immediately from its colour, and its luminosity can be deduced by measuring how bright it appears to us, then scaling upwards for the effects of distance. Obviously, the farther away the star is, the dimmer it appears. Measuring the distance can be done by estimating the quantity of interstellar dust that intervenes between Earth and Cygnus X-1. This in turn is calculated from the degree to which the starlight from the supergiant is reddened by dust absorption (just as the sun looks red through a foggy or dusty, polluted atmosphere) compared to the calculated colour for that temperature and type of star.

Although it involves a rather complex chain of reasoning, these sorts of calculations are routine for astronomers, and the result is only treated so cautiously because the issue at stake is nothing less than the existence or otherwise of the most bizarre object ever predicted by science. The upshot of the calculation is that the dark companion in Cygnus X-1 has a mass of between 8 and 18 solar masses, placing it well above the neutron star class, and comfortably in the black hole league. Nevertheless, definite confirmation of its identity will have to await more detailed observations.

Much excitement has resulted from these observations of X ray binary stars, and a great deal of theoretical effort has been poured into studying the properties of binary stars with one black hole component, using mathematical models and computer-assisted calculations. It seems likely that the system begins with a heavy star and a light star. The heavy star, for reasons already explained, evolves more rapidly than the light one, and reaches its death throes, perhaps in a supernova explosion, while the other is still burning away happily.

Before it dies, the doomed star will swell up enormously, perhaps even engulfing the lighter companion. If this happens, the lighter star will start to feed off the heavier one, sucking the tenuous material from its distended neighbour on to itself. This may continue until the lighter star actually becomes the heavier and vice versa. The core of the distended star (which shrinks as the outer layers swell) now implodes to form a black hole, and if this does not produce such a powerful supernova explosion as to disrupt the system completely, the outcome will be a fairly ordinary massive star orbiting a somewhat lighter black hole. Thus, although heavier stars evolve towards a black hole death faster than light ones, in a binary system the end product is the reverse – a heavy star and a light black hole. Sure enough, in the X ray binaries so far measured, the dead object is lighter than the living star.

The further evolution of the system, involving gases drawn from the star on to the black hole and the formation of a disc-like vortex, is not understood in detail, in spite of an immense amount of research. The structure of such a disc is very complicated – far more complicated than something like a tornado on Earth, which is still improperly understood – and may involve subtle magnetic effects as well as fluid, thermal and gravitational phenomena. Controversy rages over such issues as whether the disc is thick or thin, or will even form at all, how long it will last and whether it will tend to break up into rings or more complex arrangements. It seems as though the flow of material on to the hole will not continue for as long as a million years at the strength required to produce a strong X ray source. Probably there are no more than a few hundred such systems in the galaxy. Nevertheless, there must be many thousands that have passed through this phase during the life of the Milky Way, and are now bereft of their X ray glow, possibly with both component stars collapsed. At least one such system, called PSR1913+16, is known, although in that case the two dead stars are both believed to be neutron stars.

Both the scenarios so far discussed – the accumulation of millions of stars in a dense cluster and the implosion of the core of a single, burnt-out star – involve fairly well understood astrophysics and are amenable to direct observation. The third scenario is based entirely on mathematical theory and predicts black holes that may never be detected, but in view of its close connection with naked singularities, this theory is worth considering here.

We shall see in detail in chapter 8 that astronomers now believe the universe began with a big bang about fifteen billion years ago. During the primeval phase, the cosmic material from which the presently observed stars, planets, and so on eventually formed was exceedingly compressed. It seems reasonable that in these circumstances some of the dense matter would separate from the rest and fall together to form black holes. We know that at some stage the ordered aggregation of material occurred, because the universe is full of distinct, well-separated galaxies. In the highly compressed conditions of the primeval cosmos, one can envisage lumps of matter of all possible sizes imploding under their own irresistible weight.

Unfortunately, any attempt to analyse mathematically the number and size of these primordial black holes runs into severe difficulties connected with our ignorance about the properties of the primeval cosmic material. Opinions differ as to whether it was very stiff, very squashy or in between, and how far it started out clumped into aggregates, and if so of what size. Arguments have been advanced that suggest that supermassive holes may have emerged from the big bang and acted as the nuclei around which the galaxies subsequently grew by slow accretion of the more dissipated gases. Others have proposed that microscopic holes, containing a mass of only about a billion tonnes, could have formed in the very early stages. It is these latter objects that will turn out to have the greatest significance for the subject of naked singularities. As we shall see in the next chapter, billion-tonne microholes measure only one ten-million-millionth of a centimetre in size, which is about the same as an atomic nucleus. This means that their behaviour will be partially subject to the exotic laws that govern the behaviour of atoms – the so-called quantum theory. The outcome of this turns out to have far-reaching consequences for the nature of space and time.

In spite of their outlandish properties, black holes are taken very seriously by modern astronomers. While it would be premature to say that there is definite observational evidence of a black hole, we now know of many objects whose properties are consistent with the presence of a black hole, and for which alternative descriptions are at least no more plausible. The existence of neutron stars and quasars confronts science with the realities of ultradense, highly compact astronomical systems in which gravity has become so strong that it is seriously distorting the structure of space and time. There is no evidence for a physical mechanism that would seem to separate quasars and neutron stars from their slightly more compact cousins – the black holes. Unless our understanding of gravity and the behaviour of matter is badly misconceived, we are on the threshold already of observing systems that have imploded catastrophically to oblivion.

Given that the sorts of intense gravitational fields that will produce uncontrollable shrinkage to a black hole exist all around us in the universe, the question naturally arises as to whether these same intense gravitational fields might, having induced sudden collapse, produce not a black hole, but a singularity deprived of its event horizon – a naked singularity. In the next chapter we shall see what physicists and astronomers have discovered about the cosmic censor.

6 *The naked singularity exposed*

In the previous chapter we saw how astronomical discoveries of the 1960s and 1970s bring us face to face with a phenomenon more extreme and more outlandish than any encountered before in the history of science – catastrophic gravitational collapse. Two outcomes are predicted when collapse occurs: a black hole or a naked singularity. Black holes have bizarre properties. Nothing may escape from inside them; even information is trapped. But naked singularities open the way to still more bizarre situations, as we shall see.

The central issue is this. Consider a massive star nearing the end of its life. Its core is rotating rapidly, and therefore has a large equatorial bulge due to centrifugal force. It is buried beneath turbulent and hot material, and is about to initiate a supernova explosion, one of the most violent events ever to befall matter. The star explodes and the core implodes. As with all explosions, within the maelstrom itself there is chaos. The core is severely distorted as it collapses, and its rotation rate shoots up as it shrinks for the same reason that an ice skater spins faster by contracting the arms. The implosion of the core is therefore a highly asymmetric, turbulent and chaotic affair. What will happen? Will a black hole form, or a naked singularity?

This question is extremely hard to answer. No physicist or astronomer can hope to model the complicated situation inside a supernova in anything like the detail necessary. Even the broad principles are in dispute. All that can be done is to examine certain very idealized mathematical models of a collapsing star core and try to determine the activities of the cosmic censor. In the absence of a complete mathematical proof that the censor will always hide singularities of this sort inside event horizons, the models prove nothing if they always predict black holes, for we cannot be sure that in a more complicated or more realistic situation, naked singularities will not occur. Equally, if an idealized model does predict a naked singularity, we have to be sure that the model itself is not so idealized as to be impossible in the real world before abolishing the censor.

The two alternatives, black hole or naked singularity, are depicted schematically in a spacetime diagram in Fig. 36. In both cases a star implodes to a final singularity which is drawn on the diagrams as a wiggly line at the centre of collapse, forming at the moment that the star shrinks to nothing. In Fig. 36 (i) the distortion of the light rays folds the outgoing light into a surface that neatly clothes the singularity – the event horizon. Nothing can cross from the interior region (the black hole) outwards through this tubular surface, so the distant observer sees nothing of the events such as E, close to the singularity, as all the light from E (including the outward-directed light) falls inwards towards the singularity as shown. Certainly the observer cannot see the singularity itself.

In contrast, Fig. 36 (ii) depicts the collapse of a star to a naked singularity. The light rays (which may be extremely convoluted, for instance spiralling around the singularity) can now eventually escape to a great distance. No event horizon or black hole forms, and an observer can look directly at the singularity itself.

The first detailed study of a collapsing star was carried out, it will be recalled, by Oppenheimer and Snyder in 1939. They considered a highly idealized mathematical model in which the star was taken to be exactly spherically symmetric. The end result was a black hole of the type based on Karl Schwarzschild's original (1916) solution of Einstein's equations. In later years, other types of black holes were investigated. What happens, it was wondered, if the collapsing star is not spherical; in particular, what becomes of a rotating star? Also, suppose the star carries an electric charge, surely the electricity cannot just disappear, for the conservation of electric charge is a central law of physics?

Physicists then discovered very simple mathematical solutions of Einstein's equations which describe the exterior of black holes that rotate and carry electric charge. Quite recently, it was proved that these simple black holes are the only ones that can arise from a collapsing star, subject to some very reasonable assumptions. A lot of

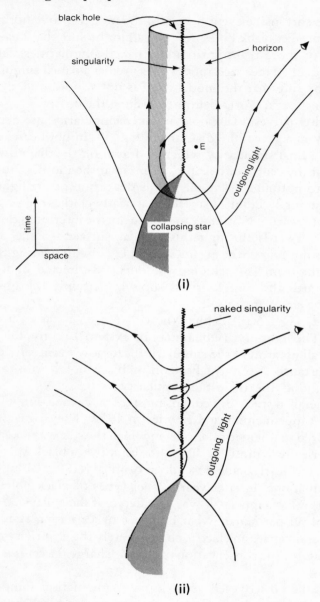

36 Black hole v. naked singularity.
(i) A star collapses to a singularity. Nearby light is prevented from escaping and is bent back by gravity. Farther away light escapes, but with difficulty. The light which just 'hovers' at a fixed distance from

the singularity forms the tubular surface, called the event horizon, that clothes the singularity, preventing events such as E from being seen by the distant observer. The region inside the horizon is therefore black – a black hole.

(ii) If a horizon does not form, light can escape – perhaps along convoluted paths, bent and twisted by the intense gravity – to be seen by the distant observer. The singularity is naked.

effort was expended in trying to understand how a complicated, asymmetric star would settle down into one of these simple black holes. It was found that during the sudden collapse, a lot of the complicated structure of the star becomes invisible from a distance, so the irregularities present at the outset do not appear in the final black hole as viewed from without. These analyses were, however, confined to small disturbances around the idealized, symmetric models of collapsing stars. Although there seems little doubt that *if* a black hole forms it will be of the simple type mentioned above, nobody knows whether, when disturbances from symmetry are great, a black hole will form at all.

In the 1960s it was discovered that the structure of the black holes that rotate or carry electric charge is quite different from the spherical, uncharged, Schwarzschild hole. Although they all contain singularities, the relation between the singularities and the paths of light rays and infalling particles is quite different. In chapter 4 it was explained how a black hole occurs: the so-called light cones are slowly tipped over by the intense gravity until the outside edge of the cone is no longer directed outwards. Inside the hole, the cones are tipped over on their sides so that all light – even outward-directed light – is dragged inwards and, in the Schwarzschild (non-rotating, uncharged) case, all particles of matter are forced towards the geometrical centre where the singularity is located.

This state of affairs is depicted schematically in Fig. 37 (i). The singularity is represented, as before, by the wiggly line. Far from the singularity, where the gravity is weak, the cones are upright, in the standard way. Then, as the gravitating centre is approached, the cones tip over until at the event horizon H (the surface of the black hole) the outside edge of the cone is vertical. Inside H, the cone is pointing in towards the singularity, and all light and matter is directed inexorably towards it. (The broken line shows the possible path of a material particle or observer.)

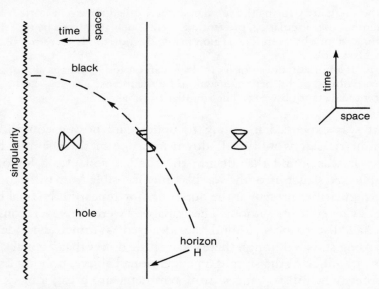

37 (i) The event horizon around a non–rotating uncharged black hole is formed when the light cones near the singularity are tipped over sideways. The direction of time v. space is also tipped over: the singularity now lies to the future. The infalling observer (broken line) cannot see the singularity till he hits it.

When an electric charge is placed on the hole, the situation alters drastically, as in Fig. 37 (ii). There is still an event horizon H, where the light cones are tipped at 45°. Inside H is a black hole from which nothing can escape. However, there is another type of horizon, marked H′, that is located inside H. Between H and H′ the light cones lie more or less on their sides, and all light and matter is forced inwards. But, when H′ is reached an extraordinary thing happens: the cones tip back upright again and the paths of particles curve back away from the singularity. Although neither light nor matter can curl right back round and out of the black hole (i.e. they cannot cross back over H) nevertheless they are no longer forced towards the singularity. It is as though there is a sort of strong repulsion around the singularity that forces infalling matter (and observers) away. A similar picture applies for a rotating black hole, but there the situation is more complicated because it is not spherical, so the direction of infall makes a difference.

If a falling observer is forced away from the singularity, but cannot escape from the hole, where does he go? It is possible to follow,

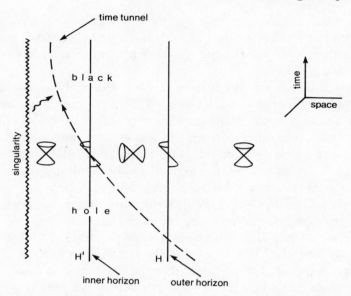

(ii) If the hole is charged or rotating, the cones tip back again near the singularity. There is therefore an inner horizon H′ as well as the usual (outer) horizon H forming the surface of the black hole. The infalling observer is forced away from the singularity along the 'time tunnel'. Light from the singularity (wiggly line) can reach him: the singularity is naked.

mathematically, where the spacetime leads, and the result is astounding. The region inside H′, around the singularity, is a sort of timelike tunnel that leads through into another universe, complete with black hole, and what could be described as a white hole from which the observer will emerge. This result was mentioned in chapter 4 where it was pointed out how an observer can fall back into the hole once more, only to re-emerge in yet another universe and so on, ad infinitum. Thus, the charged and/or rotating hole must really be viewed as an infinity of universes in such a way that departure from any one universe precludes return.

Aside from this rather absurd aspect, these black holes have another disturbing feature. In the case of the Schwarzschild hole (Fig. 37 (i)) the singularity cannot be observed, even by an observer inside it, because all the light cones are tipped over. This means that the singularity is always in the future of the observer, until he hits it. It is not possible for information or influences to leave the singularity to reach him. In contrast, the observer inside H′, the inner horizon of the rotating

113

and/or charged hole, can examine the singularity at close quarters, for it lies both in his future and past. Influences can travel out from the singularity to him. For the observer inside the inner horizon, the singularity is naked.

Is this a failure of the cosmic censor? Certainly it is true that the singularity is only naked to someone foolish enough to jump into the black hole. The rest of the universe rests safe in the knowledge that whatever horrors emerge from the singularity, they cannot escape beyond the event horizon H. The truth about the singularity is censored from, as it were, the general public. Only by special entry into the restricted 'black hole club' can this truth be witnessed. But then there is no return – club membership, and attendance, are for life.

The problem about dismissing the naked singularity in this way is that we are dealing here with questions of principle. The division of spacetime into 'the black hole' and 'the rest of the universe' is rather arbitrary. Suppose, for example, that the black hole is so big that it encompasses our entire observable universe. This idea is not completely crazy, and may actually represent an approximation to reality (see chapter 8). Even if this is not so, one can envisage (though hardly take seriously) the idea of a supertechnological community deliberately engineering the whole galaxy into a black hole. We could then quite comfortably enter the hole without at the time noticing anything terribly unusual and then, apparently, witness the singularity unclothed.

The prospect of a naked singularity inside a rotating or charged black hole has led to the general belief among physicists that, in reality, the interior of the hole would not look like the schematic diagrams shown in Fig. 37. It was mentioned above that the exteriors of all black holes are almost certainly of the simple type. However, the interior need not be simple. The structures discussed above, in particular the time tunnel to other universes and the nakedness of the singularity, are certainly present in simplified models, but it is not at all clear that they will occur in a realistic case of gravitational collapse. The reason for this concerns the nature of the inner horizon H'. It was mentioned in chapter 3 that the time elapsed as experienced by one observer need not be the same as that experienced by another observer moving differently. In particular, as we saw in chapter 4, an observer who falls into a black hole will take, in his own frame of reference, a finite time to reach it (and a very short time, perhaps much less than a second, if he falls in from near enough). But a distant observer will reckon the same sequence of events to take an infinite duration. In this sense, therefore, the interior of the black hole is, to an outside observer, beyond all infinite future time.

Examination of the time dislocation for an observer falling into a rotating or charged hole reveals an extraordinary phenomenon as he approaches the *inner* horizon, H′. During the brief duration that he takes to tumble across it, he can witness, by looking out of the hole, the whole of the history of the universe, right into the infinite future, passing by in that one fleeting moment. Everything that will ever happen is compressed into the final moment when he crosses the inner horizon. This means that all the information, and in particular all the light by which he will see the outside universe, is infinitely compressed in time. In one instant at the horizon, an infinite amount of light arrives. Aside from vaporizing him, the energy of this light will exert its own gravity, because energy, as well as matter, is a source of gravity. Being infinite, the energy will create another singularity, along the inner horizon H′, and the observer will crash into this and disappear before ever getting to see the naked singularity. Therefore, the 'naked' region inside H′, and certainly the mysterious 'other universes', can never be reached. The time tunnel is closed. In fact it doesn't exist, for the two singularities join up and destroy it. The cosmic censor wins! Of course, this conclusion depends on the assumption that there is still some light around in the universe in the very far future.

There is, however, another way that a naked singularity could appear. The size of the 'tunnel' region inside the inner horizon H′ depends on the total amount of electric charge or rotation on the hole. A small quantity produces a very narrow tunnel close to the singularity, but as more and more is added, so the region grows. A glance at the mathematics shows that if enough electric charge or rotation is supplied, the inner horizon H′ moves out so far that it reaches the outer horizon – the event horizon, H. Thereafter, any further addition of electric charge, or any increase in the rate of spin, and the horizons disappear completely, leaving the singularity exposed to the entire universe.

At first sight these considerations appear to open up the prospects of someone deliberately creating a naked singularity either by taking a black hole and dropping down more and more electric charge, or by spinning it faster and faster, say, by shooting particles obliquely into it. Alternatively, one can envisage a highly charged or rapidly rotating star collapsing to a naked singularity rather than to a black hole. This seems all the more plausible when the actual numbers are investigated. In the case of rotation, the critical spin rate at which the black hole goes over to a naked singularity is about a hundred thousand revolutions per second for a solar mass hole. That may seem stupendously fast (in fact the surface of the hole, crudely speaking, spins as fast as light) but it is not so. The sun turns once every 25 days. If it collapsed intact, by the time it reached the size of a black hole, only a few kilometres across, its spin rate (by the 'ice skater' principle) would increase a million million times. It would be spinning too fast for a black hole to form.

Does this imply that the gravitational collapse of the sun would produce a naked singularity? This is most unlikely for the following reason. As the sun shrinks and its rotation rate rises, the centrifugal force at its equator grows bigger and bigger until a point is reached where even the escalating surface gravity of the shrinking mass cannot restrain the material at the periphery. The bulging equatorial regions are simply flung off into space, taking a lot of the rotational energy with them. An examination of the mathematics shows that the condition for a black hole to form corresponds to the condition that the outward centrifugal force be less than the inward gravitational force. Therefore, the shrinking sun would continue to shed material until its rate of spin fell to precisely the value at which it would become a black hole rather than a naked singularity. The cosmic censor wins again.

A similar situation occurs with electric charge. An electric ball of matter will feel an attractive force of gravity that tries to shrink it and a repulsive force of electricity that tries to explode it. The condition on which a black hole rather than a naked singularity will form is that the electric repulsion should not exceed the gravitational attraction. Thus, to make a naked singularity by collapsing a ball of highly charged matter means shrinking a ball in which the electric repulsion exceeds the gravitational attraction. But in this case the ball will tend to expand rather than shrink. Collapse will not occur anyway.

Although the effects of rotation and electric charge are repulsive, an alternative scenario could be tried. Instead of a rotating or charged 'star' collapsing, one could start with a rotating or charged black hole, and try to increase the values of these quantities beyond the limit at which a black hole can exist. In this way it might be that the hole gets 'converted' into a naked singularity.

Suppose the hole is rotating so fast that it is very close to the limit at which the event horizon will disappear and the singularity become exposed. How could the spin rate be nudged just a little higher? One method is to drop a spinning body into the hole, along its rotation axis, so that the extra spin is imparted to the hole (see Fig. 38). This will

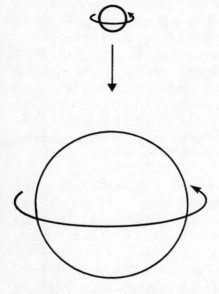

38 Spinning up a black hole. The rotating body is dropped down the spinning hole in order to increase its rotation rate. However, the extra gravity of the sacrificed body more than outweighs the rotational gain.

certainly spin up the hole, but it will also make it heavier and hence increase its gravitational grip. When this situation is investigated in detail, it turns out that the gravity increase more than compensates for the 'centrifugal' increase. The critical value of the spin necessary to produce a naked singularity goes up by more than the actual increase gained from the spinning body.

The most favourable strategy is to deposit the most spin for the least quantity of mass, as that will give the greatest boost to the centrifugal force at the least expense in extra gravity. One good choice is not to use a material body at all, but a photon of light. Photons carry spin (though very little) and also energy, which adds to the mass of the hole. However, the spin of a photon is always the same, whereas the energy decreases with wavelength. By considering photons with ever

117

greater wavelengths the energy may be reduced to as little as one pleases. This seems to offer the distinct hope of making a naked singularity, for it is now only a matter of firing a stream of very long wavelength photons down into the black hole along its rotation axis. Nature (in the guise of the cosmic censor) comes to the rescue, however. When the mathematics are examined to determine how long the wavelength of the photons must be for the hole to gain more spin than it does mass–energy (hence gravity), the answer turns out to be a surprise. The wavelength is required to be at least as big as the black hole itself. Unfortunately, a wave this big will not easily go down the hole, but tends instead to scatter off the hole and go elsewhere.

The same sort of problems afflict any attempts to charge up a black hole beyond the critical value at which the event horizon disappears. If the hole is already highly charged, any further charge will be resisted by electric force. If the extra charge is deliberately driven into the hole, such as by firing it at high velocity, the extra energy imparted to the charged particle to get it into the hole increases the hole's mass (hence gravity) by more than the charge increases the electric repulsion. Here again, the cosmic censor triumphs.

The reader may be getting the impression that a naked singularity formed from gravitational implosion of a body is just not on. This is what physicists would like to believe. Nevertheless, there are mathematical models that do suggest how naked singularities might form. One of these was discovered by E.P.T. Liang of the University of Texas. Liang considered the collapse of an infinitely long cylinder of matter surrounded by empty space. This is a system that will collapse to a naked singularity of infinite density, but not one where all the material is concentrated at a single point. Instead, the cylinder shrinks transversally, down to an infinitely long line of infinite density – a string singularity.

While this result is not in doubt, the question arises as to whether Liang's calculation is in any sense realistic. Could it be that a very long, but nevertheless finite, cylinder would not behave in the same way as the idealized model of an infinitely long cylinder? Perhaps when the effects of gravity waves coming off the collapsing cylinder are taken into account the singularity becomes clothed in an event horizon and so disappears into a black hole after all? The answers are not known, but even if they were known, there may still be a problem about drawing conclusions from an idealized mathematical model, as a close

118

analogy will illustrate.

Suppose a cylindrical pencil with a perfectly sharp point is stood on end on a horizontal surface. This is easily achieved by arranging the flat end downwards (see Fig. 39). A mathematical study of the

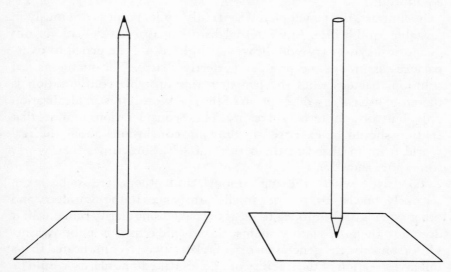

39 Stability.
   (i) The pencil standing on its blunt end is reasonably stable.
   (ii) Stood on its sharp end, the pencil will not remain vertical even though mathematically one can find a description for this condition.

problem shows this to be a possible solution. The pencil will not topple because its cylindrical shape is symmetric. There is no reason for it to topple in one direction rather than another. However, there is an alternative completely symmetric situation which corresponds to the pencil being balanced on its sharp end. Once again, there is complete symmetry, so the pencil should not topple.

In the latter case we know from common sense that the pencil will topple. Although a mathematician may predict a situation with the pencil balanced on its point, we know it cannot happen in reality (unless the point is very blunt). A more ambitious mathematical study will reveal why one solution (pencil on its base) is realistic while the other (pencil on its point) is fallacious. The reason concerns the question of stability. When sitting on its base, the pencil is quite easily toppled by a knock, but so long as the external disturbances are kept reasonably small, the pencil will remain in equilibrium. Likewise, if the pencil is not exactly cylindrical but contains chips or bumps, the equilibrium is not seriously disturbed, unless these defects are quite considerable. We say that the pencil on its flat end rests in stable equilibrium.

In contrast, the pencil placed vertically on its point is obviously in unstable equilibrium. Any external disturbance or any deviation from cylindrical symmetry will, however slight, cause the pencil to over-balance (assuming the point is perfectly sharp). The mathematical solution that describes the pencil in this unstable configuration is therefore useless as a description of the real world. It is an idealization only, with no counterpart in reality. These considerations indicate that caution should be exercised in drawing conclusions about the real world from idealized mathematical models. Sometimes they work, sometimes they don't.

To decide whether Liang's model, and others, are to be taken seriously, one has to consider small disturbances to the symmetry and other aspects of the model, to see if the naked singularity remains. If it does not, then the victory against the cosmic censor is a hollow one. Unfortunately, the general theory of relativity is so complicated that a full investigation of the stability of Liang's naked singularity would be a formidable task.

Another mathematical model of a naked singularity has been produced by Dr H. Muller zum Hagen and his colleagues at Hamburg University. They consider the case of spherical collapse, but of a slightly unusual kind. In their model, the outer layers of a collapsing ball of matter start with a more rapid implosive motion than the interior, with the result that the ball tries, as it were, to turn itself inside out. At some stage the outer layers cross and overtake the inner layers. If this can happen in a carefully defined way, a shell of infinitely dense material is produced, giving rise to a naked singularity, not at the centre of the ball, but where the layers cross.

This model has been criticized for a number of reasons. For example, the researchers assumed that the ball of matter exerts no pressure, which might be expected to disrupt the system when the density of material gets too high, although even this is not completely clear.

Another example of a mathematical model predicting a naked singularity is due to M. Demianski and J.P. Lasota. The authors consider a spherical body that collapses in the usual way, but this time light, or other radiation, flows out of the system so prolifically that it depletes the mass of the 'star' at an appreciable rate. Now the radius of a spherical black hole is proportional to its mass, so if the star loses mass, the degree of shrinkage it must undergo to become a black hole is increased. By contriving a sufficient efflux of light energy, this critical radius for the formation of the hole can be made to shrink faster than the 'star' itself. The ball of matter goes on collapsing, right down to a zero radius, without ever crossing an event horizon to form a black hole. The end product is a naked singularity, but one which has no mass left, as it has all been radiated away. We shall find a similar result in our next, and final, example.

In a black hole, the singularity is surrounded by an event horizon that prevents it being seen from without. The event horizon itself can be considered as the boundary of the black hole. The radius of the horizon is, as mentioned above, proportional to the mass of the black hole in the Schwarzschild case. If some way can be found to reduce the mass of the hole, then the hole will shrink, the horizon will contract and move closer to the singularity. If this process could be continued until all the mass has been removed from the hole, then the horizon would shrink to nothing and the singularity would be exposed.

About 1970, many physicists began investigating whether mass (hence energy) could be extracted from black holes. One such mechanism was discovered by Roger Penrose. When a black hole rotates, it sets up a sort of invisible vortex that has the effect of dragging falling bodies (and even light) around with it. Near the event horizon this effect can become very pronounced. When a body falls into a non-rotating hole there is a sense in which, on reaching the horizon, it is falling in at the speed of light. If the hole rotates, then in addition to the radial infall the particle picks up some transverse speed as well. Therefore it is possible for these two velocities – the radial infall and the sideways dragging – to combine together to give a net speed greater than that of light. The particle, which is still just outside the horizon, is travelling, in a sense, faster than light relative to an observer a great distance from the hole.

The region outside the horizon where this strange effect occurs has been dubbed 'the ergosphere', and it has some peculiar properties. It is possible for a particle to move in the ergosphere with less energy than it would have at a great distance, including the energy of its own mass.

It follows that such a particle has negative mass-energy, rather like the cavorite discussed in chapter 4. However, this negative energy is a global effect, i.e. it is a property of the whole system, and not localized on the particle. An observer in the vicinity of the particle itself would not notice anything unusual.

Nevertheless, as Penrose pointed out, this negative mass can be used to reduce the mass of the black hole and thereby extract energy from it. The way in which this could happen in practice is for an ordinary (i.e. positive mass) particle of matter to be dropped into the ergosphere. Because mass-energy is always conserved, it will retain its positive mass characteristic whilst in the ergosphere. On the other hand, if during its sojourn there the particle were to explode into two pieces, then it would be possible to arrange for one piece to fall along one of the negative mass orbits. When this piece crosses into the hole, it reduces the mass of the hole. The energy released appears on the remaining fragment, which is propelled from the ergosphere back out into space with more mass-energy than the original particle had at the outset.

At first sight it might appear that by continuing this process, the mass of the black hole could be shrunk to zero and the singularity laid bare. But once more the cosmic censor intervenes. For every quantity of mass extracted, some of the spin of the hole disappears too. This causes the ergosphere to shrink, so that the Penrose process becomes harder and harder to implement. At the end of the day the ergosphere disappears and the opportunity for further energy extraction is lost. The hole is then left with a large fraction of the original mass.

Some very general investigations were carried out by Stephen Hawking, who proved an important theorem about the amount of energy that can be extracted from black holes by any sort of process. In essence, the theorem says that whatever is done to a black hole, it will always grow larger, in the sense that the total area of the event horizon will go on increasing. Thus, although the Penrose process removes mass from the hole, it does not make the hole smaller, which was our original objective. The reason for this is that the radius of the event horizon round a rotating hole is not merely proportional to the mass as it is in the non-rotating (i.e. Schwarzschild) case, but depends in a complicated way on both the mass and rate of spin (in fact, it is not even spherical). When electric charge is added, the formula is more complicated still. During the Penrose process, both the mass and the spin rate fall, and always in such a way that the area of the hole goes up rather than down.

Hawking's area theorem applies to many other processes as well. For example, if two black holes are collided, they will fall into each other and coalesce. The theorem tells us that at the end of the process, the area of the final big black hole is always greater than the combined area of the two original holes. We cannot use black hole collisions, therefore, to make a black hole smaller.

Hawking's theorem is based on two crucial assumptions. Firstly, that naked singularities do not exist and secondly that energy and mass are always positive. The result is closely related to a very fundamental law of physics – the second law of thermodynamics – as discussed fully in my book *The Runaway Universe* – which in one form states that it is impossible to run a so-called *perpetuum mobile*. Such a device is a means of re-using degenerated energy; for example, when an electric fire has dissipated its heat into the environment, the energy is still there, but in a disordered and useless form. The second law forbids us to re-use it without expending at least as much energy elsewhere to do so. It was once fashionable for inventors to attempt to violate this law and search for a machine that would run for ever without fuel. If naked singularities exist, then it is possible for Hawking's theorem to be violated and such a *perpetuum mobile* might be possible. Naked singularities could be the ultimate answer to the energy crisis!

Turning to the other assumption behind the theorem, that negative mass or energy is impossible, one appears to be on firm foundations as far as ordinary matter or energy is concerned. However, in 1974 Hawking himself found that when subatomic matter and energy are considered, the story is spectacularly different. Hawking's main interest at that time concerned the microscopic black holes thought by some to have formed during the big bang (see the end of chapter 5). Holes with a mass of a billion tonnes (about a cubic kilometre of water) have a size comparable with the nucleus of a small atom, so that subatomic effects cannot be ignored.

The theory that treats subatomic processes is called quantum mechanics, and is explained in great detail in my book *Other Worlds*. Among other things, quantum mechanics describes processes in which subatomic particles can be created and destroyed, including photons of light. For example, quantum mechanics gives a complete description of how an excited atom may de-excite itself and emit a photon – something that goes on all the time in an electric lamp.

When quantum mechanics is applied to black holes a most remarkable and entirely unexpected result occurs. Even though there are no atoms in the hole (the matter has all shrunk away to a singularity) particles are still created, more or less out of empty space. From the region around the hole stream electrons, protons, mesons, photons, neutrinos – in fact, every conceivable type of subatomic particle and antiparticle. More extraordinary than the appearance of these subatomic particles is the fact that their energies correspond precisely to those that would have emerged from a body that had been brought into thermal equilibrium with its surroundings. This implies that there is a characteristic temperature associated with the hole, and in the case of the mini-holes it is not inconsiderable – about a hundred billion degrees. Evidently black holes are not black after all, and mini-holes are white hot.

The discovery that small black holes radiate like furnaces brought a whole new dimension to the question of naked singularities and Hawking's area theorem. If the hole radiates energy in the form of subatomic particles, it must pay for it somehow by losing mass. But how? The particles themselves do not come out of the hole directly, because nothing can get out of a black hole. Closer examination reveals that the whole notion of the location of a subatomic particle becomes meaningless anyway over such small dimensions.

After further investigation it became clear how the black hole was losing mass. The peculiar properties of quantum matter and energy enable negative mass-energy to appear in very restricted locations. Around the black hole this negative energy is produced by the intense gravity, and it streams across the event horizon into the hole. The mass of the hole thereby falls, not because matter has escaped, but because negative mass has flowed in. As a consequence of the flow of negative mass into the hole, the conditions necessary for Hawking's area theorem to apply are violated, and the area of the black hole decreases. In short, as the black hole radiates heat energy into its environment, it slowly shrinks in size.

One of the oddities about the quantum black hole is that the temperature of the hole goes up as the mass goes down. This means that for a big hole, such as one caused by the collapse of a star, the temperature is minute, around one ten-millionth of a degree above absolute zero. These black holes are very, very nearly completely black, and their tiny quantum radiance would be utterly undetectable. The mini-holes, however, are a different proposition. At a hundred billion degrees they would radiate fiercely into their surroundings and lose heat energy at a prodigious rate. Moreover, as they lose energy and therefore mass, their temperatures go up, in contrast to ordinary systems that cool as they lose heat. It follows that the radiation process is unstable, and escalates, with the black hole getting hotter and hotter, and dispensing heat ever faster. As the process continues, so the black hole gradually evaporates away.

In contrast to the almost instantaneous collapse of matter to form a black hole in the first place, the quantum evaporation is leisurely in the extreme. In spite of the enormous temperature of the hole, its total mass–energy reserves are colossal. Our billion tonne hole has enough energy to power the world's total consumption for hundreds of millions of years – and it is all locked up in something smaller than an atomic nucleus! It takes billions of years before the quantum radiation has any drastic effect. However, when the energy reserves finally run out, the result is dramatic. The temperature starts to shoot up beyond any value of which science has direct experience, and the rate of evaporation spirals out of control. The black hole has reached a great crisis. The horizon starts to shrink noticeably in years, then days, then seconds, then microseconds. As the final reserves of mass dwindle, the evaporation of the hole turns into an explosion, and a great burst of energy is released in one go. The black hole itself has, apparently, shrunk away to nothing, the event horizon having been squeezed down on to the singularity. All the mass has been extracted from the hole to leave . . . what? A naked singularity? If the black hole evaporation process is taken at face value, it does indeed seem to lead inevitably to a massless naked singularity. Just what such an object would look like or do nobody knows, though there is much speculation (see next chapter).

The evaporation of black holes to naked singularities seems to be based on much former ground than the other suggestions for producing naked singularities made earlier in this chapter. This is partly because the evaporation will occur for all black holes, irrespective of whether they have rotation or electric charge; indeed, one effect of the radiation is that it reduces both these quantities. Therefore, one does not encounter the objection that the result is based on the sort of mathematical idealization discussed earlier, analogous to pencils balanced on their tips. If an idealized black hole evaporates, so does a disturbed one. A second reason for having confidence in Hawking's result is that the quality of the radiation emitted by the hole is very special, i.e. that associated with a body in thermal equilibrium. It has already been mentioned that the second law of thermodynamics has a close analogue in Hawking's area theorem. Now we find that quantum black holes behave just like thermal bodies. This seems more than just a coincidence, and suggests a deep link between black holes and thermodynamics. For that reason one is inclined to take the black hole evaporation process seriously.

This issue of thermal equilibrium is so crucial that it is worth special consideration. During the winter of 1859–1860, the German physicist Gustav Kirchhoff directed attention to the following property of radiant heat. If a lump of material is suddenly heated in one place, some heat will tend to spread throughout the material, while some will be radiated away into the environment from the material's surface. Because parts of the body and the environment are hot, and parts are not quite so hot, the radiant heat will be spread among many different wavelengths in a haphazard way. The hotter parts will tend to radiate shorter wavelengths than the cooler parts.

If, instead of a lump of material, one examines a cavity completely enclosed by walls that are shielded to prevent heat escaping to the surroundings, then after a while the walls of the cavity will reach a uniform temperature as the heat spreads evenly throughout the material. Some heat radiated inwards, into the cavity, will be reabsorbed by the opposite face of the cavity, so will not disturb the thermal equilibrium that maintains the entire system at a uniform temperature.

Examination of a patch of wall surface inside the cavity will show that it radiates heat energy at exactly the same rate as it absorbs radiant heat coming from the opposite wall. This must be so, otherwise equilibrium would not be sustained. Consider, therefore, a beam of heat radiation with one particular wavelength. When it strikes this patch of surface some of the heat radiation will be absorbed and some will be scattered, or reflected, back into the cavity. The precise proportions of absorbed and reflected heat will depend on the nature of the material that lines the walls. For example, metal is highly reflective, whereas black paint absorbs heat readily. In spite of this uncertainty, whatever proportion the material does choose to absorb, it must still be exactly equal to the quantity of heat that the same patch of surface radiates into the cavity in the same interval, or the balance between emission and absorption that characterizes thermal equilibrium would be upset.

It follows from this simple argument that the amount of radiant heat energy present in the cavity at any particular wavelength remains unaltered. But because the cavity is at a uniform temperature, and because the heat radiation bounces around in a fairly random sort of way, there must be the same amount of radiant heat everywhere throughout the cavity, even including regions where the walls have different reflective properties. (One could, for example, paint black patches on a metal surface.) Kirchhoff argued that if the amount of heat is the same in the vicinity of all materials, then it must be independent of whatever material the cavity walls are made of. That is, the energy of radiant heat at any given wavelength must be dependent only on the temperature of the system, and not on the type of cavity.

Kirchhoff's reasoning depends crucially on the assumption of thermal equilibrium. It tells us that however complicated the distribution of heat energy among the various wavelengths will be for a glowing lump of substance, when it is in equilibrium with its surroundings it will always radiate the same spectrum of heat for that temperature. If some cavity material completely absorbs all the heat radiation that falls on it then it will, to maintain equilibrium, radiate precisely this same characteristic spectrum. For that reason, the very special quality radiation emitted by a body in thermal equilibrium is called 'black body' radiation. Thus, Hawking's black holes radiate exactly like Kirchhoff's black bodies. It is an intriguing connection.

If Hawking is correct, then can we expect that even now, out in space, mini-black holes are exploding into naked singularities? Calculations show that the energy released in the last tenth of a second of its vanishing act amounts to the equivalent of a million megaton bomb, which sounds impressive, but by astronomical standards is an inconspicuous flash. Much of the energy would be released in the form of high-energy gamma rays. Satellite-born gamma ray detectors ('telescopes') exist, and some have actually detected bursts of gamma rays, though none with the characteristics expected from black hole explosions.

Further analysis indicates formidable problems with attempts to detect black holes using gamma ray telescopes. There can scarcely be more than a few mini-holes per billion billion billion cubic kilometres, or they would contribute more mass to the universe than the galaxies; we should then notice their gravitational effects. An even more stringent restriction is provided by the fact that the accumulation of gamma rays from all the black hole explosions going on throughout the universe has failed to register on the equipment. It turns out that even in a cubic light year of space, there cannot be more than a few dozen of these objects. Thus, even taking an optimistic estimate, the nearest explosion likely to occur during one month's observing time is about ten light years away, which is hopelessly far for detecting a modest explosion of mere nuclear proportions.

In spite of these pessimistic estimates, there is a possibility, pointed out by Martin Rees of Cambridge University, that secondary effects from the explosions would be detectable at a very great distance. The hole, while in its final moments, reaches such a high temperature that not only gamma rays are produced. All sorts of subatomic fragments – electrons, positrons, protons, neutrons, mesons, and so on – will also vomit forth. Many of these microscopic particles are electrically charged, opening up the possibility of powerful electromagnetic disturbances from the outburst of debris. In particular, a hole which explodes in our cosmic neighbourhood will be surrounded by the weak magnetic field of the Milky Way galaxy. When the shower of electric particles explodes out into the ambient magnetic field, the field disturbance created rumbles off as an electromagnetic wave, most noticeably a radio wave in the frequency band 100 to 1000 MHz. Because radio telescopes are so much more sensitive than gamma ray telescopes, and also have the advantage of being Earth-based, a modest search using existing equipment might be capable of detecting black hole explosions anywhere in the galaxy. If Rees is correct in this idea, then the fact that radio astronomers have not discovered exploding black holes implies already that there are unlikely to be more than one mini-hole explosion per million cubic light years per year. With dedicated effort, even a density one millionth of this figure would be enough to be detectable.

Although the search for primordial mini-holes is daunting, the pay-off is considerable. Not only would one of the most dramatic scientific predictions of the century be confirmed – in a subject where there are precious few experimental restraints on the scientific imagination – but our knowledge of the primeval cosmos would be enhanced enormously. The exploding mini-holes represent a relic from the earliest imaginable cosmic epoch that we can realistically expect to probe (see chapter 8).

In addition to securing a powerful observational underpinning for an exciting area of gravity physics and cosmology, the exploding holes would provide a unique opportunity for physicists to study directly the behaviour of ultra-energetic matter, at temperatures unlikely ever to be attained by other means. Even in the world's most powerful subatomic accelerators, the energies released correspond to a temperature of only about a million billion degrees. During the last few years of its existence, an evaporating black hole will be hotter than this, and its temperature will be rising fast. The sort of subatomic particles that emerge at these colossal temperatures will have been absent from the universe since its origin. A study of them could prove of immeasurable value in understanding the fundamental forces of nature.

The evaporating black hole and the absence of a definite proof of cosmic censorship suggest that the naked singularity is a serious proposition. If that is so – and many physicists would be very reluctant to accept it – then nature is threatened with anarchy. When a singularity bursts upon the universe, the rational organization of the cosmos is faced with disintegration.

7 Facing the unknowable

The desperate crisis that would befall the physical universe, if a naked singularity were to appear, can best be appreciated by first understanding the nature of cause and effect, and the role of determinism and prediction in science.

In primitive societies long ago, people had almost no comprehension of the natural phenomena that occurred around them. There were regular, dependable events, such as the phases of the moon, the eclipses, and the seasons. There were also sudden, violent outbursts, like earthquakes and flooding. All these happenings were attributed to the whims and fancies of the gods, and were supposed to be directly caused by supernatural intervention.

In later centuries, with the establishment of the scientific method based on careful observation and experiment, together with mathematical analysis, it became clear that the explanation for many of these familiar phenomena rested, not with divine causes, but with other physical phenomena. The eclipses, for example, were explained as the geometrical alignment of the Earth, sun and moon caused by their regular motion around each other in mathematically well-defined orbits that could be computed on the basis of Newton's laws of gravity and motion. Volcanic eruptions or sudden earth tremors were found to be caused by the build-up of pressure and stresses in the Earth's crust.

Today, the greater part of the world about us is understood in this way. Things happen because other things happen, and they in turn happen because of preceding events, and so on, in a chain of cause and effect that ultimately encompasses the entire universe. Thus, the

130

build-up of stresses in the Earth's crust is in turn caused by the forces in the Earth's interior that slowly drive the plates into which the terrestrial surface is divided. These driving forces are caused, at least in part, by the heat flowing from the Earth's core, and this heat may be attributed to the release of energy in radioactive decay occurring in the heavy elements located in the core. The radioactive processes are themselves understood in detail, and are caused by instabilities in the competing forces that bind atomic nuclei together. And so on.

Of course, there are still gaps in our knowledge and understanding of the workings of the physical world. Biological systems, for instance, are only poorly understood. Yet few scientists believe that any events occur without a prior physical cause which is, given enough research effort, describable in detail in terms of some mathematical theory. These days even theologians are disinclined to believe that God ever directly interferes in the running of the universe, with the possible exception of the creation.

The belief that every event that occurs in the universe has its origin in some preceding causes conjures up the image of the physical world as a complex network of influences, closely interwoven, and continually acting and reacting between the component parts, with a multitude of forces and fields. Although at present only a fragment of the total picture is understood, it is easy to suppose that, in principle at least, everything that ever happens in the universe is completely determined by everything else.

Pierre Laplace once remarked that, if he had complete knowledge of the state of the universe at one instant – if he knew where every atom was located and how it was moving – then he could in principle, using mathematics, predict the entire future of the universe in detail. The basis of this extraordinary claim was the belief that everything which will happen, say, tomorrow is caused by (and hence determined in detail by) all the things that happen today. The reasoning is obvious: if no extraneous influences, such as divine intervention, invade the universe, then the cause of every one of tomorrow's events must have its origin somewhere among today's events. Or so it seemed to Laplace. We shall see below that things are not quite as simple as was supposed by the philosophers and scientists of previous centuries.

A crucial feature in these ideas of determinism is that our would-be predictor must have complete knowledge of the entire universe. This is necessary because, if any region of the universe were omitted, influences could emanate from that region and invade the part of the universe of interest, thereby spoiling the prediction. For example, the motion of the planets around the sun can be worked out in complete detail purely from a knowledge of the solar system itself, but the predicted future behaviour would be badly wrong if a stray planet flung from some distant part of the galaxy were to career unexpectedly through the solar system.

With the discovery of the theory of relativity in 1905, the above ideas had to be modified somewhat to take into account the fact that no physical influences can propagate between bodies faster than the speed of light. This means that it is, after all, permissible to ignore distant parts of the universe when making a prediction of the future behaviour of a restricted region, provided the prediction is limited to moments that are not too far in the future. Thus, in the case of the solar system, one may predict one year ahead with knowledge restricted to events occurring within one light year of the solar system's periphery. The stray planet that threatens to rush in on the scene must be at most one light year away if it is to arrive here, travelling at less than the speed of light, within one year. We should therefore be obliged to include this object in our domain of causation one light year in radius, and account for its influence on impending events. Nothing that happens outside one light year can ever affect what occurs in the solar system within the next year. Naturally, the further ahead in time one wishes to predict, the greater the volume of the universe one must consider in order to be sure of catching all the causative influences.

The central role played by light rays in the subject of causality leads to complications when the volume of space involves extragalactic distances, for then one must take into account the expansion of the universe and the curvature of spacetime. This can have a profound effect on the nature of prediction and the lawful behaviour of the cosmos. To illustrate the possibilities we shall study in detail the properties of a model of the universe first proposed by the Dutch astronomer Willem de Sitter in 1917. It is based on Einstein's general theory of relativity, with its description of gravity as curved spacetime.

De Sitter's model universe is best understood with the help of a picture. Chapter 1 introduced the idea of a spacetime map in which time is drawn vertically and space horizontally. To represent both three-dimensional space as well as time on a sheet of paper means suppressing at least one space dimension, and if we also wish to depict the curvature of spacetime, only a single dimension of space can be accommodated. Spacetime will therefore be represented as a sheet,

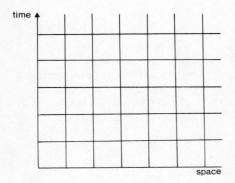

40 (i) Representing spacetime as a sheet, one can imagine parallel lines of 'longitude' and 'latitude' (called coordinates) to label the location of events.

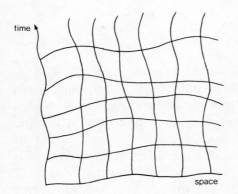

(ii) When gravity is present, everywhere parallel coordinates are impossible, as spacetime becomes distorted.

which can be curved by the effects of gravity. Figure 40 shows the sort of universe that can result when gravity crumples up spacetime; the lines are only included for aid of visualization.

If the spacetime sheet can be curved, there clearly exists the possibility that the sheet can be joined up with itself in a closed shape such as the cylinder shown in Fig. 41. Thus, we cannot be sure that the topology of the universe is the same as that of an infinitely extended sheet. It may be that it more closely resembles the cylinder. What does this mean physically?

Remembering that time runs vertically and space horizontally, the cylinder spacetime of Fig. 41 represents a world in which space does

133

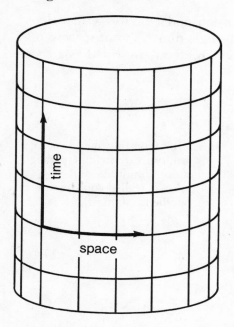

time

space

41 Cylinder universe. This figure represents a universe in which space is finite in volume. The cylinder should be envisaged as infinitely extended vertically.

not extend for ever, but curves around and joins up with itself. If we wish to take a 'snapshot' of the entire universe at one instant, this is represented by a horizontal slice through the cylinder, which yields a circle. The circle is a one-dimensional attempt to represent a finite, closed, three-dimensional space. In such a space, a traveller can set off in a fixed direction and pass right around the universe, returning to his starting-point from the opposite direction, like a cosmic Magellan. Just as the circumference of the circle is finite, so the volume of this cylinder universe is also finite; one could, in principle, visit every place in the universe. Notice, though, that although this universe is finite in volume, there are no barriers or boundaries, no centre and no edge. Every point on the circle is equivalent to every other point.

The idea that space may not extend to infinity, but that there may still be no edge to the universe appears bewildering, and cannot be properly envisaged by thinking in terms of models such as the diagram shown in Fig. 41. Nevertheless, it makes good physical and geometrical sense, and can be checked by astronomical observation. Unfortunately it will be a few more years yet before astronomers have telescopes powerful enough to decide definitely whether space continues to infinity, or is closed and finite like the cylinder.

One question that always crops up in the discussion of closed space is the meaning to be attached to, for example, the region inside and outside the cylinder. It is important to remember that spacetime is represented only by the two–dimensional sheet itself, the region inside the cylinder is not part of the physical universe. The only reason for considering it is as an aid to visualizing the topology of the cylinder, but it is not logically necessary to include it.

To appreciate this point, imagine the experiences of an observer permanently confined to the sheet, just as we are permanently confined to spacetime. The observer travels round and round the universe at constant speed, returning again and again to his starting-point. The track of such an observer is a helix, shown in Fig. 42. The intervals of

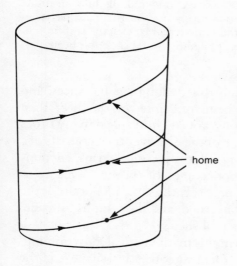

42 An astronaut repeatedly circumnavigates the finite space of his cylinder universe.

home

time between revisiting any particular place, such as 'home', are always the same. Our observer cannot see into the regions inside or outside the cylinder, and even has some intellectual difficulty in understanding the idea of an 'inside' or 'outside'. He knows his world is finite in volume, but he need not model it by appealing to a higher-dimensional curvature as we have done. Instead he could continue to think of his world as an infinitely extended flat sheet, consisting of an endless sequence of vertical strips (see Fig. 43), each identical.

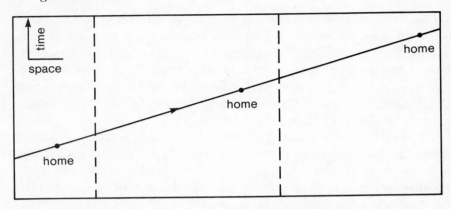

43 The experiences of the astronaut discussed in Fig. 42 could be
reinterpreted without the need to invoke a cylinder at all. This infinite
sheet is divided vertically into identical strips. As the astronaut passes
through each strip he experiences the same features, exactly as if he
were circumnavigating the cylinder. (The horizontal scale has been
increased somewhat.)

If one imagines the cylinder being rolled along the flat sheet, and
each of its features (such as 'home') being imprinted on the sheet, then
one obtains a map of the cylinder 'unwrapped', multiplied many times
as it rolls round and round. The track of the observer then consists of a
straight line passing in sequence through the identical strips, encountering again and again exact carbon copies of 'home'. This two-
dimensional multi-strip universe is precisely the same physically as the
cylinder so far as the observer is concerned, and the regions 'outside'
and 'inside' the cylinder have been abolished.

It is not always possible to perform this mathematical 'unwrapping'
manoeuvre in higher dimensions or for any geometrical shape (e.g. the
sphere) but the physical considerations remain the same. Although the
space around the spacetime sheet is helpful for visualization, as far as
an observer restricted to the sheet is concerned, it is purely fictitious
and need not be invoked at all. The geometrical and topological
properties of the space make perfect sense in terms of observations
confined entirely to within the spacetime sheet. And the same is true
for real three-dimensional space.

A four-dimensional cylinder universe was invented by Einstein in
1917. However, it does not give a very good picture of the real
universe, even if it is established that space is closed and finite. The

reason for this is the discovery made by the American astronomer Edwin Hubble in the late 1920s that the universe is not static, but in a state of expansion. Hubble discovered that all galaxies are moving away from all others in a systematic pattern of recession.

The expanding universe is now a cornerstone of cosmology, but its nature is often misunderstood. Many people (including some scientists) think of the recession of the galaxies as due to the explosion of a lump of matter into a pre-existing void, with the galaxies as fragments rushing through space. This is quite wrong. There is no astronomical evidence that the universe has a centre or an edge, as the idea of an exploding lump would imply. Moreover, if space is finite, it seems that the galaxies would be distributed uniformly around the space and not concentrated in a blob.

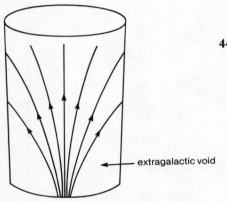

extragalactic void

44 Big bang – wrong picture. A huge primeval lump explodes and the fragments fan out to fill the available space in the cylinder universe.

The popular, but erroneous, arrangement is depicted in Fig. 44. The world lines of galaxies fan out towards the future, as the whole assembly of galaxies disperses to fill eventually the available volume of space. In contrast, the correct picture is shown in Fig. 45. Here the galaxies do not disperse at all through the space. What happens instead is that the galaxies at all times populate the space uniformly (i.e., they are distributed evenly everywhere with no central concentration), but the *cylinder itself* expands. The expanding universe is not, therefore, the motion of galaxies *through* space, away from some centre, but is the steady expansion of space.

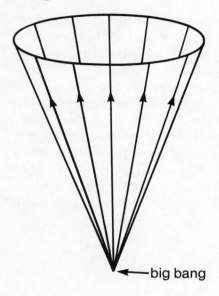

←big bang

45 Big bang – right picture. The galaxies do not move, but the space expands from nothing. The intergalactic gaps swell with time, reproducing the effect of the galaxies moving apart.

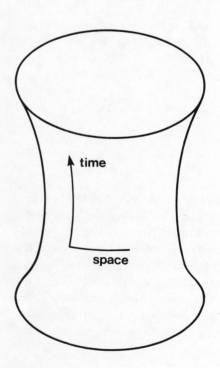

time

space

46 De Sitter's model of the universe. This hyperboloid-shaped sheet represents a finite space that contracts to a minimum volume, then expands out again at an accelerating rate. In this figure and those that follow, the sheet should be envisaged as infinitely extended vertically.

In the real universe, we can think of the cosmic expansion as due to a progressive swelling of space everywhere. Around each galaxy, every day, no less than a hundred billion billion billion billion billion cubic kilometres of new space appears from nowhere, as it were. The distance between the galaxies increases, therefore, not because they are moving apart in the traditional sense, but because the available space is continually swelling.

De Sitter's model universe incorporates this cosmic expansion feature, even though its was proposed several years before Hubble's great discovery. The shape of de Sitter spacetime is shown in Fig. 46, and it will immediately be noticed that although the top half represents an expanding space, the bottom half is a reflection of this, so it represents a contracting space. Thus, in de Sitter's cosmos the universe contracts from an infinite volume, bounces at some minimum size, and expands outwards again. Another important feature is the way the spacetime 'sheet' curves outwards at top and bottom. This means that the expansion gets progressively faster as time goes on. We shall see below what implications this has for the propagation of light and the nature of causality in the de Sitter model universe.

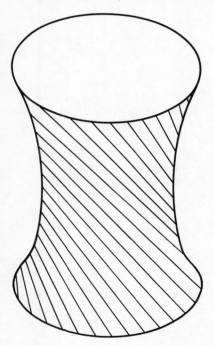

47 Basket weave seat. The hyperboloid surface of Fig. 46 can be constructed from a bundle of straight rods inclined in the way indicated. In de Sitter space, these lines could be the paths of light rays. Similar rays (not shown) would slope from bottom left to top right.

The technical name for the shape of the surface shown in Fig. 46 is a hyperboloid. If it is sliced horizontally, the sections are circles, representing the closed, finite space. If it is sliced vertically, the lines are hyperbolae, each representing the entire history of one galaxy (i.e., its world line). In spite of these rather technical features, the hyperboloid is a familiar shape, because it is possible to make it entirely out of straight rods. It may seem rather surprising that a surface that curves in two perpendicular directions can be made from straight rods, but it is this property that endears it to furniture manufacturers, and hyperboloidal seats made from basket weave are quite common. The arrangement of rods is shown in Fig. 47.

In the de Sitter model universe, the straight lines that occupy the positions of the rods in the basketware stools could represent the paths of light rays in space. Figure 48 shows two pulses of light emitted in

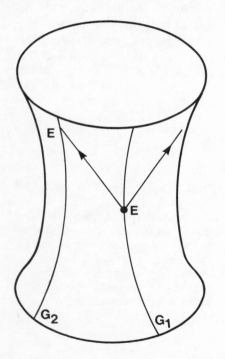

48 At some moment E in the history of galaxy G_1 a flash of light is emitted. The light travels in all directions (only two can be accommodated on our diagram) along the straight 'rod' lines of Fig. 47. On reaching galaxy G_2 at event E', the light is red-shifted because G_1 and G_2 are receding on account of the expansion of their universe.

opposite directions at some event E along the world line of some galaxy. The light rays, therefore, are straight lines, even though the surface in which they are embedded is curved.

Suppose we consider two neighbouring galaxies G_1 and G_2 (see Fig. 48). Light signals can be sent between them; indeed, this is how inhabitants of one can see the other. But as the de Sitter universe expands, so the two galaxies move apart faster and faster. This recession shifts the wavelength of the light that passes between them because as the space stretches, so the light waves stretch with it. Thus, the received light is somewhat reddened by the expansion. (Red light has a long wavelength.) In fact, it was this characteristic redenning in the light from distant galaxies that led Hubble to infer that the universe is expanding.

The farther apart the galaxies get, the faster they recede and the redder the received light becomes. Modern telescopes can detect galaxies billions of light years away that are receding at a good fraction of the speed of light, and the wavelength of their light is stretched to several times its natural value. Eventually, the two galaxies G_1 and G_2 would be so far apart that the light would be shifted completely beyond the visible spectrum, and neither galaxy could be seen by inhabitants of the other. When the rate of recession exceeds the speed of light itself, the wavelength is stretched infinitely and no communication of any kind is possible. The situation is therefore reminiscent of the black hole, where communication ceases at the event horizon because of an infinite red shift.

Another view of the event horizon in de Sitter space is obtained by considering the geometry of the hyperboloid. Suppose at one particular instant E_1, a light signal is sent out from galaxy G_1 to nearby galaxy G_2 (see Fig. 49). On receipt at G_2 the signal, which is somewhat red shifted, is bounced back to G_1. It arrives back at galaxy G_1 some time later, at event E_3, by which time the galaxies have been swept a lot farther apart and are now receding rather fast. The return signal is therefore considerably red shifted. If this signal is then bounced back again to G_2, Fig. 48 reveals that it will never arrive. The light ray goes right 'over the top' of G_2's world line and never intersects it. It is always beyond G_2's infinite future. It follows that, however long one waits, the event E_3 will never be able to influence what happens to G_2. In de Sitter's universe, therefore, a complete prediction of the entire future behaviour of a system only requires knowledge of what is happening at one instant in a restricted patch of that universe. Events such as E_3, which are beyond G_2's event horizon, can be ignored. In the same way, events inside a black hole can be ignored as far as what goes on outside is concerned, for they can never exert any influence outside the hole's event horizon.

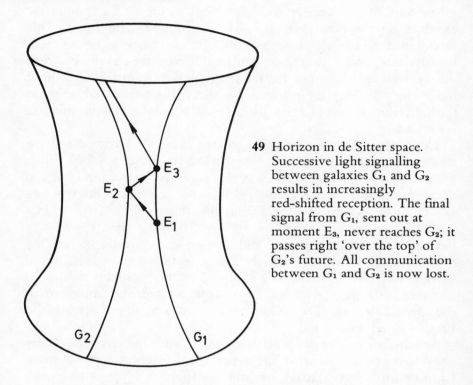

49 Horizon in de Sitter space. Successive light signalling between galaxies G_1 and G_2 results in increasingly red-shifted reception. The final signal from G_1, sent out at moment E_3, never reaches G_2; it passes right 'over the top' of G_2's future. All communication between G_1 and G_2 is now lost.

It appears that de Sitter's universe is even safer from 'outside influences' than the one that Laplace had in mind. The real problems with cause and effect come when one considers not de Sitter space, but a related model universe in which the hyperboloid shown in Fig. 46 is tipped on its side, so that time runs in a circle around the closed surface (of which more later), while space is represented by the infinite horizontal hyperbolic slices (see Fig. 50). Thus, space in this model, which is sometimes referred to as anti–de Sitter space, is infinite in extent, unlike in the case of de Sitter space considered above.

In anti–de Sitter space the 'rods' still play the role of light rays, but their relation to the galactic world lines is different because these now run round the throat of the sheet. Clearly, every light ray will somewhere intersect every galaxy, because the world lines of the galaxies wrap right around the surface. There is thus no event horizon.

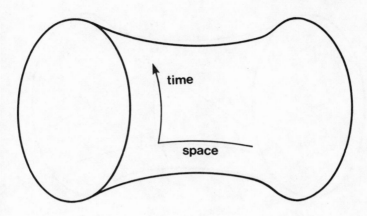

50 Anti-de Sitter space. This model
universe is similar in shape to de
Sitter space, but with the
hyperboloid on its side. Space
(horizontal slices) is now
infinite, but time (vertical slices)
is circular, i.e. finite in duration.

The surprise comes when one considers the whole of space at one
instant of time, which is represented by a horizontal slice. The
hyperbolic line which results will extend out to infinity in both
directions (left and right in the diagram), but in spite of this there will
still be some light rays that do not intersect this hyperbolic line. In de
Sitter space, the hyperbolae are galactic world lines, and the circles
represent space at one instant. There, it is the world lines that 'miss'
some light rays, and this leads to an event horizon. In anti-de Sitter
space, the roles of space and time are interchanged, so it is now the
spatial slices that 'miss' the light rays.

What does this mean physically? Consider a galaxy G (see Fig. 51)
and the state of the entire infinite universe at one moment, simultaneous with event E_1 on G's world line. In Laplace's model of the
universe, that information would have been sufficient to predict the
entire future of G, for the universe at this instant would contain all the
causative influences that could ever reach G. But in anti-de Sitter
spacetime things are very different. The spatial slice curves away
below the light rays marked L, so that, even if the state of every atom

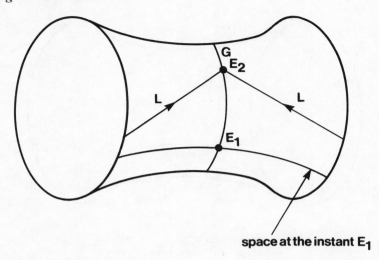

space at the instant E₁

51 Collapse of cause and effect. The event E_2 (e.g. vaporization of an observer by pulses of light L) does not owe its origin to any event whatever, anywhere in the whole of space, at the earlier time that the observer reckons to be simultaneous with E_1. The origin of the light pulses must be sought from 'beyond infinite space'.

and every light pulse were known at time E_1, it would be insufficient to predict what would happen to the galaxy G after event E_2, when the light rays L, and their later associates, bring in influences from 'beyond infinite space'.

To make these rather abstract ideas more concrete, suppose that G contains a paranoid observer who accumulates every scrap of information about the entire universe to assess any threat to himself. Even with complete knowledge of every atom at that moment (which is in any case impossible to obtain directly, as he has to wait for this information to reach him at the speed of light) he would still not be able to ensure that at some later time, a pulse of light might not streak in from space and vaporize him.

This bizarre form of spacetime, where influences can appear literally out of nowhere – from beyond infinity – enjoys a sort of divine intervention. Rather than seeing the behaviour of every object as having its causative origin in some earlier event, we now have the possibility of things happening that cannot be traced back to any event *that has yet happened*. The chain of cause and effect is broken. We shall see that this is also a feature of naked singularities.

144

Before leaving anti–de Sitter spacetime, a few remarks must be made about the peculiar nature of time in this model. As remarked above, the world lines of galaxies wrap around the sheet and could even join up with themselves as circles. This means that time is closed, and the past is also the future, a situation that was briefly considered in chapter 3. Such an arrangement seems to lead to causal chaos; it would, for example, be possible to know the future by keeping records of the past, even though the future is unpredictable from a complete knowledge of the present!

Returning to the subject of naked singularities, it is easy to see that they share a number of unpalatable features associated with anti–de Sitter space. Figure 36 (ii) shows a spacetime diagram in which a star collapses to form, not a black hole, but a naked singularity. Assuming that we do not live in something like an anti–de Sitter universe, the behaviour of the star, right up to the moment that the singularity forms, is completely and precisely determined by the condition of, and events in, the universe at prior moments. After the singularity has formed, however, disaster strikes. All of physics breaks down at the singularity itself, so it is not possible to know what influences may emerge from it. We cannot predict what will happen to a singularity.

Just as influences flow into anti–de Sitter space from the 'sides' of spacetime (i.e., 'beyond infinity'), so when a naked singularity is present, uncontrollable influences can flow out from it, i.e. from across the edge of spacetime. A singularity is a boundary to physical space – a place where totally unpredictable influences can invade the universe. But whereas in de Sitter space these influences stream in from beyond the infinite, here they stream out from a localized place.

The most disturbing thing about a naked singularity is that it seems to mark the end of science and the orderly operation of the universe. Here is a thing that is not subject to any laws of nature – at which all laws are smashed by the rip in spacetime – lying naked, able to influence what goes on around it. Surely this is a recipe for cosmic anarchy?

We have seen that in the past, humans invoked gods to cause the unpredictable, and viewed the world as a temperamental place, full of caprice and random occurrences. Then, with the growth of science, nature came to be regarded as lawful and the universe to be organized according to strict mathematical principles. Now, with the threat of a naked singularity, we are brought back once more to the chaos of the early days – to a universe in which anything at all can happen.

What can one say about a universe that contains a naked singularity? What would it be like? This question is, in a fundamental sense, unanswerable by definition, because in such a world nothing is predictable. We can, however, envisage a number of scenarios. First, there is the totally chaotic singularity. In this case, the influences that flow into the universe through the singularity are completely disordered. Matter and energy emerge from it at random, showering the surrounding universe. As ordered material, such as stars and planets, drops into the singularity, sucked down by its immense gravity, out flow unstructured atoms and radiation, rather like a permanently exploding bomb. The net effect of this shower of energy is to create a sort of background noise in which the behaviour of other objects in the vicinity is disturbed in an unpredictable but unsensational way. Moreover, the effects of the noise will diminish with distance from the singularity. Such an object, which represents the most benign type of naked singularity, could easily exist elsewhere in the universe without our being aware of it.

The next scenario supposes that the influences that emerge are partially organized. One can imagine an Alice-in-Wonderland world in which the singularity coughs out all manner of weird and wonderful pre-formed objects – stars, planets, people, micro-electronic apparatus, copies of *Encyclopaedia Britannica*. In a lawless universe, anything goes. There is no doubt that we should regard such an object in rather the same way as our ancestors regarded the Deity. It would, after all, display a sort of intelligent behaviour, and be capable of interfering in the running of the universe in an organized way. It would also play the role of a limited creator, taking in any odd debris that falls towards it and spewing out structured and ordered objects.

A third scenario elevates the role of the singularity from that of a sort of local handyman to chief global architect. In this picture, the singularity takes over control of the universe in a big way, and furnishes more than the occasional TV set. Instead, it acts as a complete cosmic rejuvenation mechanism.

In my book *The Runaway Universe*, I have explained how the entire cosmos is slowly but surely disintegrating, as all the organized structure and elaborate activity around us gradually runs down. This inexorable death of the universe has been known for a century and is a consequence of the so-called second law of thermodynamics which, in its most general form, requires that in every natural process the total degree of disorder in the universe increases. Examples of this tendency abound: people grow old, houses fall down, mountains become eroded, stars burn out. Of course, there are examples of systems that become progressively more ordered, such as social organization, but only at the expense of a greater degree of disorder elsewhere (e.g. the depletion of natural resources). In all cases, when a balance sheet is drawn, disorder wins. The entire universe is declining irreversibly towards total chaos.

The most conspicuous degeneration of the cosmos is the steady depletion of fuel in stars. Stars supply the free energy that drives most of the noticeable ordered activity around us. The sun, for example, powers the biosphere on Earth, as well as the weather and other processes such as atmospheric changes. As the sun uses up fuel it hastens the day when it will burn out and become a cold, compact object. For more massive stars the ultimate stage of this stellar depletion is a black hole, the most disordered object known, for nothing at all survives an encounter with it. When an object drops down a black hole, all information about it is wiped out for ever.

These rather depressing considerations have led most scientists to assume that, in the unimaginably distant future, the universe will be a burnt-out husk, devoid of use or interest. All that, however, could be changed by a naked singularity. Out of this object could come all the organizing necessary to wind the universe up again and to keep it ticking over.

It is amusing to consider in detail how this might happen. The naked singularity would have to act rather like a recycling device, whereby the material of old, burnt-out stars slowly migrates towards it, falls into it and disappears from spacetime, to be replaced by fresh hydrogen gas spilling out into the surrounding cosmos to form new stars that would then pass through their own life cycles. Obviously, a balance would have to be set up so that the rate of infall and outflow were equal on average.

It is not clear how this complicated arrangement might work in practice, as it is necessary for material both to fall in and emanate from the same object. It seems unlikely that the system would work in an expanding universe. In spite of this, a few years ago the respected cosmologist George Ellis proposed a model universe that contains a naked singularity as a recycling mechanism, which he claims gives almost as good a description of the real universe as the conventional model.

The Ellis universe is rather like the cylinder universe considered earlier in this chapter, except that the Earth is located on one side and a naked singularity on the other. There is no cosmic expansion. Instead the galaxies are arranged very unevenly, with a great deal of material crowded round the singularity, and very little near the Earth. The effect of such a distribution of matter is to produce a red shift of light that, at the Earth, has the same characteristics as if the galaxies were receding.

The explanation for the red shift is easy to understand. In chapter 3 it was explained how light can tire as it climbs up against gravity, and becomes redder in the process. In the conventional model of the universe this does not happen to light travelling between galaxies, because the galaxies are fairly uniformly distributed, and so the gravitational forces from different directions cancel each other out. There is no preferred 'up' and 'down'. In the Ellis universe, by contrast, the matter is clustered round the singularity, so light which leaves this region experiences a strong resisting force due to the gravity of the accumulated masses. It therefore emerges red shifted. Conversely, light which leaves Earth and falls towards the singularity is *blue* shifted – it gains energy as it plunges down towards the gravitating centre. There is thus an asymmetrical relationship between the region of space in the vicinity of the Earth, and what we could regard as our cosmic antipodes – the concentration of matter round the singularity, located on the opposite side of the universe. The galaxies crowded into its vicinity all seem red to us, but the Milky Way would seem blue to an observer on one of them.

The reason why this type of non-uniform distribution of galaxies is not usually invoked to explain the cosmological red shift discovered by Hubble is that it requires the Earth to be located at a very special place. We know from observation that the galaxies around us, in every direction, all appear red. In the Ellis universe this symmetry only occurs in the region of the antipodal point to the singularity, i.e. at one special location. Ever since Copernicus demoted the Earth from the centre of all creation, it has been anathema to scientists to assume that our location in the cosmos is in any sense privileged. In the standard picture of the expanding universe described earlier in this chapter, *all* galaxies see their neighbours red shifted, because each moves away from every other. Therefore, the Milky Way (our galaxy) need not have any special status in order to witness red shifts in all directions.

To 'explain' why the Milky Way just happens to be located near the antipodal point of his model universe, Ellis points out that the physical conditions necessary for the existence of life are only to be found well away from the singularity, which is very hot. Hence it is no surprise that we, as intelligent observers, find ourselves living in this particular region of the cosmos – it is the only region that is habitable.

Ellis conjectures that heavy elements, such as iron, that represent the ashes of stellar nuclear fuel, somehow get spewed into space (for example in supernovae explosions) and slowly migrate round to the naked singularity where they disappear from existence, to be replaced by fresh hydrogen. In this way, the universe achieves infinite longevity, with the singularity acting rather like a permanent and inexhaustible life support system.

Of course, this type of model universe is not intended to be taken too seriously. When naked singularities are involved, one is free to speculate that anything is possible. The singularity could in principle control everything that ever happens in the universe, right down to the tiniest detail.

If naked singularities really can exist, how should we face up to the ultimate unknowable? If one were to form near the Earth tomorrow, would the world descend into a madhouse, a lawless assortment of random uncaused events, so that we should never know from one moment to the next what would happen? Would forces be unleashed that would destroy the universe or endow it with unlimited life?

Only a handful of scientists have studied the nature of singularities in detail and there is no agreed answer to questions of this sort. Hawking has argued that, being an utterly lawless entity, a singularity should originate totally chaotic and random influences. It might then be very little different from an object like a quasar, which is highly compact, extremely energetic and pouring out vast quantities of disordered energy. In that case, a naked singularity, well away from the Earth, might be of no more consequence to us than all the other apparently random influences that come from a great distance. It must be remembered that even if one could in principle predict the future, in practice no one could ever have enough information about the distant galaxies actually to do so. In the real world we just have to take our chances anyway on whatever the universe throws at us. In this respect, a chaotic naked singularity is no different from what we are used to.

The situation can be compared with an electronic system over which the listener has no control. It might play anything at all. An ordered system may produce a Mozart concerto, a partially ordered system fragments of speech, and a disordered system would produce white noise. Hawking envisages a singularity as a white noise generator, not of sound, but of light, heat, subatomic particles, and every other type of material and radiative influence.

In a sense, total chaos is like a law, for if we know the naked singularity produces white noise, we know something about its average, statistical behaviour – the complete absence of correlations. We need not worry about it producing an identical copy of the reader, or a lump of antimatter to annihilate us, for these require a degree of co-operation and organization. In this way, the naked singularity is tamed. We face it squarely, and instead of hoping such things never occur, we treat them as a part of nature – a source of white noise. Viewed in this way, they become rather innocuous entities.

How can we tell what might come out of a naked singularity? Is Hawking right that they are completely disordered systems, bathing the universe in random and not especially threatening influences, or is Ellis' extraordinary idea correct, with a naked singularity injecting order into the cosmos in a continual revitalization? Or do we expect the madhouse outcome, with all sorts of quirks and oddities invading the world around it?

Fortunately, we do have some information, limited though it is, about what might come out of a singularity. If recent astronomical discoveries are to be believed, one such has occurred already. In the next chapter we shall see that the universe itself came out of a naked singularity.

8 The creation of the universe

Belief in a creation of some sort has always been prevalent in most Western cultures. The plain fact that man can multiply to fill empty territory must have made it obvious that, in the past, there were fewer people, so it would have seemed natural to suppose there had been a first man and woman from whom all humanity derived. As humans were long regarded as the pinnacle and purpose of the physical universe, early cultures had little difficulty in presuming that the remainder of the universe had been created along with, or immediately prior to, man himself.

Today we have direct scientific evidence that man has only inhabited the Earth for a few million years, though we now know that there was never a 'first man' in the old religious sense. Evolution of complex bioforms from simpler predecessors can be traced in detail through the fossil records back to over three billion years ago. In addition, radioactivity analyses have established the age of the Earth at about 4½ billion years, while the sun is thought to be a little older. Before this, the solar system did not exist.

How many other features of the universe that we now know did not exist at some remote epoch in the past? Many of the other stars in our region of the Milky Way galaxy are known to be of comparable age to the sun. Since the second world war, astronomers have gained a detailed understanding of how stars are born, evolve through a definite life cycle, and eventually die. There are many places in the sky, some of which are visible to the naked eye, where great clouds of hydrogen gas are slowly contracting under gravity and fragmenting to form huge glowing balls, shrouded in nebulous filaments, that will become the next generation of stars. It is probable that the sun formed this way, amid a whole cluster of other stars that eventually dispersed among the galaxy, some 5 billion years ago.

A couple of hundred years ago scientists had no real grasp of the concept of energy that has recently assumed such fundamental importance in our lives. It seemed perfectly natural that the sun and stars should go on pouring out heat and light ad infinitum, without the need for a source. We now know that this huge energy output must be paid for somehow, and with the discovery of nuclear processes the power source of the sun and stars was explained. It therefore became clear that, as with all fuels, the nuclear fuel that powers the stars will eventually run out. The subsequent fate of the depleted star was described in detail in chapter 5.

The fact that the stellar fuel is inexhaustible obviously implies that the stars cannot always have existed, and a measure of the rate of fuel consumption together with a knowledge of the nuclear reactions that release the energy reveals that even the oldest stars are unlikely to be much more than ten billion years old. By piecing together the results of radioactivity studies with an analysis of stellar evolution, astronomers have dated the Milky Way galaxy to around ten or twelve billion years old. This seems to be typical of other galaxies as well.

This figure of somewhat over ten billion years is a fascinating result. In the previous chapter it was explained how the astronomer Edwin Hubble had, in the late 1920s, discovered that the galaxies are engaging in a systematic recession from each other, a phenomenon readily explained in terms of Einstein's theory of relativity as the swelling or expansion of intergalactic space. By measuring the speed of recession of different galaxies at various distances, it is a simple matter to deduce how fast the universe is expanding. At the present rate of expansion, the observed universe will be twice its present size in about twenty or thirty billion years. Conversely, ten billion years ago, when the galaxies were born, it would have been substantially smaller than now.

It is obvious that if the galaxies are moving apart now they must have been closer together in the past, but how close? Can we be sure that the present rate of galactic recession is typical of all cosmic epochs, or was the universe expanding faster or slower in the past?

There are two ways in which astronomers have approached these questions. The first is by direct observation. Light travels at 300,000 kilometres per second. Although exceedingly fast by human standards, the scale of astronomical distances is so great that light takes an enormous length of time to travel between galaxies. For example, one of our nearby neighbouring galaxies, Andromeda, is nearly two million light years away, which is to say that light takes two million years to reach us from Andromeda. In other words, when we look at Andromeda, we see it as it was two million years ago. At the other end of the scale, the distant astronomical objects known as quasars are several billion light years distant, and the light by which we see them today left those objects before the Earth even existed.

Because of the light–travel delay, a telescope is also a timescope, enabling astronomers to look back into the universe of long ago. Among other things that they can examine is the rate of cosmic expansion in the remote past, to see if it differs at all from the present rate. In principle this is a straightforward enough task, but there are technical complications. The very distant objects – which are the ones seen in the oldest light – are also the faintest, and therefore easy to miss. There is a natural tendency to pick out only uncharacteristically bright ones. The problem here is that the only way of knowing just how far away a very distant galaxy or quasar may be is by observing how bright it appears. Obviously, a luminous object of fixed intrinsic brightness looks fainter the farther away it is located: a car headlight ten miles away cannot rival a candle in the same room. Unfortunately, not all galaxies (and certainly not all quasars) have the same intrinsic brightness; some are many times more luminous than others, even of a similar size and shape. If a random search selects a higher proportion of bright ones, these being the easy ones to spot, there will be a tendency to underestimate their true distance from Earth.

In addition to the distortions introduced by selection affects, there is much uncertainty about how the intrinsic brightness of galaxies changes with time. It cannot be assumed that the average galaxy was as luminous in the past as it is today, so this too will complicate the distance estimate, and hence the computation of the date at which astronomers are really viewing the object. In spite of all these problems there has been (with the exception of some recent dissent) general agreement that the rate of cosmic expansion is slowly and steadily diminishing – the universe is slowing down. The rate of deceleration is still too uncertain, however, to predict confidently whether the deceleration is sufficient to bring the expansion to a stop at some stage in the far future.

A decelerating expansion is precisely what one would expect from the second method by which astronomers have studied the cosmic motion, i.e. using mathematical modelling based on the theory of gravity. As explained in chapter 4, gravity is a purely attractive force (except perhaps under very exotic circumstances). It follows that the gravitational forces that act between galaxies will try to pull them together, in the same way that the material of each individual galaxy is held together. As the galaxies move apart, so the intergalactic gravitational forces restrain their motions, continually sapping the energy of expansion. The expansion motion should therefore slow up with time. It is rather like a rocket that blasts away from Earth into space. The payload leaves the vicinity of the Earth travelling at very high speed, but as it climbs away into deep space, the Earth tries to drag it back, and in so doing gradually slows down its rate of recession, perhaps even snatching the payload back altogether.

In addition to this, gravity is known to diminish with distance, according to the 'inverse square law' discussed in chapter 1. This means that in the past when the galaxies were closer together, their mutual gravitational attraction would have been much greater than now, and the rate of deceleration of the expansion still more pronounced. Thus, gravity resisted the dispersal of the galaxies much more vigorously several billion years ago than today. It follows that the expansion rate itself must have been much higher then in order for the galaxies to have escaped each other's gravitational grip and avoid falling together.

Taking into account the fact that the expansion rate of the universe has been steadily falling, it is clear that, say, twelve billion years ago, when the galaxies were forming, the universe would have been very much more shrunken than our previous estimate, based on the present expansion rate, would suggest. In fact, most astronomers agree that some time between twelve and eighteen billion years ago, the universe was so shrunken that individual galaxies were not discernible at all, being all squashed together with no intergalactic spaces. The fact that the date at which the galaxies began to separate, estimated from measuring the rate of cosmic expansion, coincides with the ages of the galaxies, as computed from radioactivity and stellar evolution studies, is remarkable confirmation that these basic cosmological ideas are correct. There was a time, about fifteen billion years ago, when there were no galaxies.

We arrive at a vivid picture of the early universe, possessing none of the empty spaces so characteristic of the present arrangement. Instead, the whole cosmos was filled with gas, more or less uniformly distributed. As the rapid expansion of space proceeded, so 'cracks' began to appear in the gas, and the precursors of the galaxies began to separate out. Eventually, the blobs of separated gas contracted under their own gravity to form the galaxies that we see today. The 'cracks' opened up into chasms, and are now vast tracts of emptiness, even wider than the galaxies themselves. The majority of the universe is now intergalactic void, containing at most only exceedingly tenuous gas.

What can be said about the state of the cosmos that preceded the appearance of the galaxies? From what has been said, it is clear that the pre-galactic gases must have been expanding very much more rapidly than now. It is a simple matter to explore backwards in time and compute what the size of a given volume of the presently observed universe would have been at any given moment, and to estimate the rate of expansion it must have had to avoid gravitational collapse.

A convenient unit to work with is the total volume of the universe currently accessible with modern telescopes. This is about a thousand billion billion billion cubic light years, one cubic light year being several thousand billion billion billion billion cubic kilometres. In this volume of space are located many billions of galaxies, each containing on average a hundred billion stars of mass comparable to the sun. The total mass of material is therefore around a hundred thousand billion billion billion billion tonnes, with an average density of one gram of matter per million billion cubic kilometres, or one atom per thousand litres.

Let us start by going back many billions of years before the separation of the galaxies, to a time when the presently observed universe was contained in a volume about one ten-billionth of the value quoted above, and the expansion rate was a hundred thousand times faster than now; the universe was doubling in size every few hundred thousand years. As we pass back in time, only one hundred thousand years before this the same material was compressed into a volume thirty thousand times smaller still, and the expansion rate was up at a hundred million times its present rate.

Clearly the pace of change accelerates as we probe farther back. At a mere one hundred years before this, the whole of the presently observed universe was squeezed into a spherical volume only about 30,000 light years across – which is the volume of space currently occupied by the Milky Way galaxy. If we push back to just one year before this time, we encounter the extraordinary image of all that we now see – the billions of galaxies, the 100,000,000,000,000,000,000, 000,000,000,000,000,000,000,000,000 tonnes of matter – crammed into a volume no larger than ten light years across and expanding so fast that its size doubles in under a second.

Following the mathematical progression, the change accelerates ever more rapidly. Passing back to one second before this drives up the density of matter to nuclear values and beyond, until the whole of the presently observed universe is crushed into a volume the size of a bucket, with the expansion rate a veritable explosion, doubling the size in a mere thousand-billion-billion-billionth of a second.

How far can we go back? The mathematical formula tells us that for ever smaller steps backward, the compression rises ever faster, and the rate of change was ever more frenetic. Following the mathematics to its ultimate conclusion, the moment is reached when everything is concentrated into the same place, the universe shrunk to a mathematical point, and the present enormous volume accessible to our telescopes was squeezed to zero, its material contents compressed to an infinite density. The situation is reminiscent of the collapse of a star inside a black hole, but played in reverse, for this is an expansion. It looks strongly as though we shall encounter another singularity.

Before tackling the central question of whether the universe has expanded out of a singularity, it is important to assess what observational evidence there is to support the assumption that there was accelerated compression back in the era before the galaxies formed. What do we know about the primeval epochs that preceded the images available in our telescope-timescopes?

If it is true that the universe started out expanding explosively fast, we might expect the early stages to have been marked by extreme violence. This expectation has led to the name 'big bang' for these events. The sort of energies unleashed by a bang of cosmic proportions would presumably have generated a lot of heat. Are there any relics left over from the primeval universe that show evidence of extreme heat?

In 1965 two employees of the Bell Telephone Company were

experimenting with satellite communication when they discovered a mysterious background of very short wavelength radio waves, called microwaves. It soon became clear that the disturbance was emanating from outer space, and subsequent measurements showed that the mystery radiation has an energy spectrum very close to what one might obtain from a body that has come into thermal equilibrium (i.e. a black body) at a temperature of three degress above absolute zero (−270°C). Most astronomers now believe that the microwave background radiation from space is none other than the primeval heat radiation itself, cooled now to the very depths, but still bathing the entire universe in its dwindling glow. Good evidence to support this interpretation comes from the fact that the radiation is equally strong in all directions of space, ruling out any 'local' origin within our galaxy, for such would be expected to show large directional variations due to the uneven distribution of galactic matter around us. Clearly the radiation comes in from deep extragalactic space. Yet we know there is not enough matter there to produce such a vast quantity of radiation until one traces it right back to the highly dense, compressed conditions of the big bang – before there were any intergalactic spaces.

Further evidence for a hot big bang comes from an examination of nuclear processes that can be driven by the huge quantity of primeval heat. Just as the compression can be computed at any given time from a mathematical progression, so can the temperature. For this purpose it is more convenient to express the result in terms of time elapsed since the beginning of the explosion, i.e. the moment of infinite compression. At that moment the temperature would apparently have been infinite too. By one second it had fallen to ten billion degrees, and by a few minutes the temperature was down to a few per cent of this. These are the sorts of temperature encountered in the centres of the hottest stars, and one would expect prolific nuclear reactions to take place.

Before one second, the temperature was so high that ordinary atomic nuclei could not have existed. This means that the primeval cosmic material consisted of a soup of individual subatomic particles roaming about in a chaotic mêlée. The primary constituents were protons, neutrons, electrons, photons (heat and light) and neutrinos. The main action concerns the protons and neutrons. A nucleus, such as that of the carbon atom, consists of a ball of neutrons and protons glued together by powerful nuclear forces. The simplest such union is that of one neutron and one proton, forming the nucleus of what is known as deuterium, otherwise called heavy hydrogen (this is chemically identical to hydrogen but about twice the weight). Somewhat more complex, but far more stable, is the further union of two deuterium nuclei (two neutrons and two protons) to form the nucleus of the helium atom. Continuing in this way the list passes on through lithium and beryllium to carbon (six neutrons, six protons) and so on up to uranium (about 150 neutrons and 92 protons).

In the primeval universe, once the temperature had fallen below a few billion degrees, the way was open for the fusion of the protons and neutrons into these heavier, more complex, composite nuclei. However, the time available for nuclear reactions was limited for three reasons. The first was the plummeting temperature, which after a short while fell too *low* for the reactions to proceed efficiently. Secondly, as the universe expanded, the ever swelling volume of space had to be populated by a roughly fixed number of protons and neutrons. Thus the density of these particles rapidly dwindled, thereby steadily reducing the likelihood of close encounters between them. Thirdly, neutrons do not remain neutrons for ever outside the confines of the nucleus. On their own, unbound by the nuclear force, they disintegrate after about a quarter of an hour.

Calculations show that during the few minutes after the first seconds of the big bang, about one sixth of the protons and all the neutrons got bundled up into helium nuclei, with a tiny fraction left halfway as deuterium, and an even smaller fraction of heavier elements such as lithium and carbon. The remaining primeval material consisted of the residue of unaffected protons. These individual protons eventually became the nuclei of the lighest and simplest element – hydrogen. Therefore, if the hot big bang theory is correct, one would expect the universe now to contain about 25 per cent helium, a trace of deuterium, and nearly all the rest hydrogen. This picture will, of course, have been distorted somewhat because of the subsequent production of heavier elements that has occurred since then in the centres of stars (see chapter 5). However, the broad picture of a universe consisting almost entirely of hydrogen and helium in a ratio by weight of about 4:1 is predicted. Remarkably, for a theory so economical on assumptions, this prediction is strikingly accurate.

Accepting then that the hot big bang model is a close approximation to the real primeval cosmos, our attention is irresistibly directed to that first instant when the universe went bang. At this stage it is worth recalling the point made in the previous chapter that the big bang was not the explosion of a lump of matter into a pre-existing void, but the sudden, explosive appearance of space and matter out of nothing. This idea is so alien to common sense that its full meaning must first be explored. As explained in chapter 7, the expanding universe is not the dispersal of galaxies away from some common centre of explosion, but the inflation of space itself. We can envisage, picturesquely, little globules of space busily breeding, and elbowing each other out of the way to make room for the progeny. Back in the big bang the breeding rate was progressively faster the nearer one probes the initial instant. And just as the ancients envisaged the entire human race springing from one man, so we can think of the entire observable universe descending from one infinitesimally small globule of space.

Before continuing along these lines, a further common misconception must be dispelled. It was mentioned in the previous chapter that space might be either infinite in extent, or curved upon itself in a finite, yet unbounded higher-dimensional 'sphere'. In the latter case, one can really suppose that the entire universe began compressed into one point. On the other hand, if space is infinite, we have the mathematically delicate issue of conflicting infinities, because infinitely extended space becomes infinitely compressed at the beginning of the big bang. This means that any given *finite* volume of the present universe, however large one chooses it to be, was compressed to a single point at the beginning. Nevertheless, it would not be correct to say *all* the universe was at one place then, for there is no way that a space with the topology of a point can suddenly assume the topology of a space with infinite extent. For this reason, the terminology 'observable universe' rather than 'whole universe' has been used in the foregoing discussions.

If the above model of the big bang is taken seriously, and the mathematical progression pushed right back to infinite density and zero volume, then we cannot continue back beyond that point. Whenever infinity is reached in physics, the theory stops. Taken literally, space has disappeared, along with all matter. Whatever lies beyond, it does not contain any places, or any things in the usual sense of material entities.

We seem here to be on the very edge of existence once again, at a singularity similar to the end point of a collapsing star, but in reverse – in our past. It is this feature that distinguishes the big bang singularity from the black hole case – the former is naked. There is, however, an important distinction between the nakedness of the big bang singularity and that of the hypothetical object depicted in Fig. 36 (ii) and discussed in the accompanying paragraphs.

Because spacetime cannot be continued through a singularity, the singularity itself can, as has been already remarked, be regarded as a sort of edge or boundary to spacetime. If a spacetime diagram is drawn, with time running vertically and space horizontally as usual, then the singularity can be thought of as a border to the diagram. Crudely speaking, the fabric of spacetime has been cut through there to yield an open edge. The cut can either be made roughly horizontally or roughly vertically (or any combination, or angle in between) across the diagram. The properties of the singularity depend crucially on which is the case.

A roughly vertical cut is called a timelike singularity because it lies mainly along the line of increasing time, whereas a roughly horizontal cut is called spacelike. The former can be regarded as a boundary to space, whereas the latter is a boundary to time. If the horizontal cut is at the top of the diagram then it is a boundary to the future; at the bottom, the cut bounds the past. The big bang singularity is a past boundary to time – a spacelike singularity that implies a past temporal edge of existence.

The significance of the difference between timelike and spacelike naked singularities is that the latter, big bang, type of singularity does not endure with time. It is, in a sense, instantaneous. Whatever unfathomable quirks a spacelike naked singularity gives us, it gives them all in one go. A timelike naked singularity on the other hand continues with time and can go on and on infesting the universe with new influences.

These differences are crucial when it comes to questions of cause and effect, and the concepts of predictability and lawfulness of the physical world. In the previous chapter it was explained how the presence of a naked singularity destroys science by allowing things to happen without a lawful cause – indeed, anything may happen in the vicinity of such a singularity. These remarks remain true in the big bang case, but they are in one sense less serious: the lawlessness only characterizes the instant of the big bang, the very beginning of the explosion. Thereafter, law and order take over and the universe unfolds according to well-defined, disciplined principles.

The existence of a past edge to time implies that, in a very fundamental sense, the big bang represents the creation of the physical universe. It is not just the creation of matter, as so often envisaged in the past, but a creation of everything that is physically relevant, including space and time. In the traditional religious picture, space and time are always regarded as absolute – transcending the affairs of matter, and existing for ever. Now we see that time may not continue into the past or future for ever. The big bang was the beginning of time. Whether there will be an end of time for the whole universe is still an open question.

We can now view the creation as a special case of a naked singularity. Anything can come out of a naked singularity – in the case of the big bang the universe came out. Its creation represents the instantaneous suspension of physical laws, the sudden, abrupt flash of lawlessness that allowed something to come out of nothing. It represents a true miracle – transcending physical principles – that could only occur again in the presence of another naked singularity.

The existence of a naked singularity, and hence a creation, at the beginning of the big bang is based on a very simplified model of the universe that supplies a mathematical progression, given the size of some typical volume of space – say, the presently observable universe – at any given time. We must now question some of the simplifying assumptions made here, in particular the precise meaning to be attached to the word 'size'. As already remarked, the galaxies are very uniformly distributed throughout space. Moreover, the expansion of the universe is equally fast in all directions. This regularity means that a given geometrical shape – for example, a sphere of galaxies – will remain a sphere as the expansion proceeds. It will not distort to a cigar or pancake shape as it would if the expansion were more rapid in some directions than others. Thus, 'size' can refer to the volume of a sphere of matter without worrying about 'shape' also. Only the size changes, the shape remains the same.

Although the assumption that the shape of a given volume of space does not change with time is clearly a good approximation, it can only be approximate. There are certainly local irregularities in the organization of cosmic matter, as the galaxies themselves testify to a certain degree of non-uniformity and clumpiness. Just as the collapse of an exactly spherical star to form a singularity inside a black hole is an idealization, so too is the assumption of an exactly uniform universe emerging from a big bang singularity.

As remarked in chapter 4, it was once thought that departures from exact symmetry could prevent the formation of a singularity, but the Hawking–Penrose theorems proved that it was not so. Even if most of the matter misses it, a singularity of some sort will always form if gravity remains attractive and a trapped surface exists somewhere in the spacetime. In the same way, the big bang singularity cannot be avoided by appealing to departures from exact symmetry. However, it might well be that the nature of the singularity is much more complicated than in the idealized case. Instead of the observable universe emerging in an orderly way from a single point, it could have erupted from a hideously distorted, twisted, extended region of infinite density.

Consider, for example, a spherical volume of space encompassing the presently observable universe. The arrangement of galaxies within that sphere will not be precisely uniform. One would certainly expect a slightly more rapid expansion in one particular direction than others, though we know from observation that the differences are now very small. Nevertheless, as we run the cosmic evolution backward towards the creation, it is likely that these small differences were progressively magnified until, back near the beginning, our spherical volume was stretched out like a cigar, or perhaps squashed like a pancake, or some still more convoluted shape. It would then have been possible for the universe to be at an infinite density without all the matter being at the same place. As explained in chapter 6, it is only necessary for it to be compressed into an infinitely thin wafer, or extended along an infinitely thin line.

One possibility for a complicated big bang singularity is depicted schematically in Fig. 52. As well as being convoluted in 'shape' (i.e. in space) it is also 'corrugated' in time, showing peaks protruding into the far future of the universe. The singularity is partly spacelike and partly timelike. An observer at the event P could therefore examine at close quarters a fragment of the creation, and even travel *into* the creation (along the world line shown). In this case, the big bang singularity is not an instantaneous affair, and one could envisage lots of little 'mini-bangs' going off even now, representing the peaks of the corrugations in Fig. 52. Indeed, there have been attempts to identify

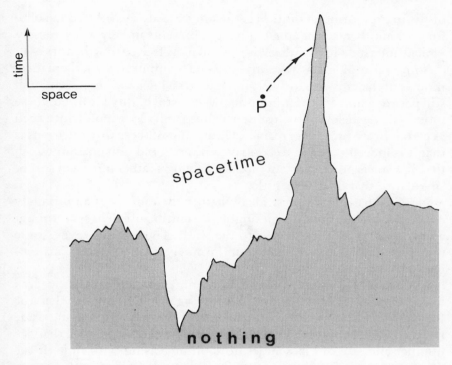

52 Big bang and little bangs. The creation of the universe could have been
a convoluted naked singularity that must be regarded as a past
boundary or edge of time. The peaks are sort of delayed creation
events – little bangs. An event P is shown that lies to the future of the
big bang, but in the past of a little bang, so an observer may travel
from P (broken line) to the edge ('into the creation').

these delayed creation events with the explosive outbursts seen from
the cores of some galaxies and from quasars. Although both naked and
enduring, the corrugated big bang singularity still does not seem quite
so dreadful as the naked singularities (discussed in the previous
chapter) that can be formed at will by appropriately manipulating the
collapse of ordinary matter.

It is undoubtedly the case that, in the real universe, the big bang
singularity was an immensely complicated and tangled affair. The
question is whether its convoluted shape left any 'holes' through
which spacetime could continue without coming to an edge. Could it
be that, although there was a singularity of some sort in our past, it did
not create *all* of spacetime, but only some of it? Did it only create some

of the matter, or none of it? Has the universe always existed in some form or another, so that the big bang represents merely an extremely violent interlude in an otherwise singularity-free cosmic history?

Some cosmologists have investigated complicated mathematical models of the universe to ascertain whether a big bang singularity will still permit a universe of infinite age. While still highly idealized, these models nevertheless allow for departures from exact uniformity such as certainly are present in reality. The upshot of these investigations is that it is indeed possible to contrive infinitely old universes in which the big bang is just a transient phase involving a rather impotent sort of singularity that does very little.

Nobody has the slightest idea whether the models of an infinitely old cosmos or the ones with a singularity cutting off all of spacetime at a universal creation are nearer to the truth. There is, however, some rather general evidence against an infinitely old universe. This evidence concerns black holes.

In chapter 3 it was explained how, once matter has disappeared inside a black hole, it cannot get out again. This means that the formation of black holes is an irreversible process and, leaving aside for the moment the Hawking evaporation effect, it implies that the number and size of black holes goes on increasing with time. If the universe were infinitely old, it might be expected on general grounds that nearly all matter would have disappeared into black holes by now, and the universe would be totally different from what is observed.

This simple conclusion is unfortunately complicated by a number of considerations. If the universe which expanded out of the big bang existed before the big bang then it can only have been contracting prior to that epoch. It cannot have remained static and then exploded, because there would have been no way in which the material could avoid falling together under gravity. If the universe was always contracting before the big bang then, as we look farther and farther back towards the infinite past, the cosmic material would have been infinitely dispersed. The question then arises as to whether it was clumped together in galaxies or smaller units, or spread uniformly, perhaps as individual atoms. In the former case, it is hard to see how, in the infinite time available, all the material in each unit could have avoided falling together to form black holes, which would then have evaporated after an immense period of time, leaving a universe almost entirely composed of radiation, with very little matter. It is possible for a pre-big bang phase dominated by this type of radiation to be made consistent with the presently observed universe, by postulating the conversion of some of the radiation back into matter during the big bang.

This conjecture runs into the added difficulty that the newly created matter would (unless certain recent speculative theories about sub-atomic forces are correct) be accompanied by an equal quantity of so-called antimatter. In itself, there is nothing objectionable about the antimatter, except than on contact with ordinary matter both species are completely annihilated back to radiation again. Thus, a mechanism has to be found to separate the matter from the antimatter, perhaps into distinct aggregates encompassing whole clusters of galaxies. Though we cannot tell whether other galaxies are made of matter or antimatter, it is known that very little mixing of the two takes place, or the resulting gamma rays produced during their annihilation would be detectable. While some cosmologists have devised theories that go some way to explaining how matter and antimatter can be separated, there is no really convincing mechanism known. Thus, attempts to model the universe along these lines get rather bogged down in technicalities. Of course, one could simply postulate *ad hoc* that the big bang singularity coughs out the right amount of matter to give us the presently observed ratio of matter to radiation, but this cannot consti-tute a proper explanation.

If the material in the pre-big bang contracting universe began dispersed as individual atoms, then the collapse into black holes could have been averted until the density of material rose high enough for local gravitational disturbances to become important. This would imply that for an infinite time the universe quietly contracted until, as the big bang was approached, the material began clumping together into galaxies, stars and a few black holes. The universe then reached a state of maximum compression, destroying all the stars (but not the black holes), whereupon it 'bounced' in some way that we do not understand, and exploded out again in the big bang. Naturally, some of the matter, and presumably some of the black holes, would have had to avoid the inevitable singularity that we know must have been present. While this general scenario cannot be ruled out, it in some ways requires an origin of the universe even more remarkable than a sudden, singular creation of everything. Can we really believe that, throughout the universe in the infinite past, all the individual atoms

165

were neatly spaced out so that, even in the immense time available, no large gravitating bodies would form?

There is, in fact, another scenario for a universe of infinite age. It has already been remarked that the force of gravity acting between the galaxies operates to restrain their recession, causing the dispersal rate to diminish with time. If the universe contains sufficient matter, the restraining force will be strong enough to halt completely the cosmic expansion and turn it into a collapse, dragging the galaxies back on to each other in a sort of big crunch – the time reverse of a big bang. The Hawking–Penrose singularity theorems predict that a singularity will occur at the end of the collapse, and most cosmologists assume that this would mark the annihilation of the entire universe – a future end of space, time and matter. This can be represented on our spacetime diagram by a horizontal cut along the top. A universe that has both past and future spacelike singularities only exists for a finite time, say, one hundred billion years. Indeed, time itself would be only a hundred billion years in duration.

If we wish to speculate that the universe existed before the big bang, then we could suppose that it will also survive the big crunch, and bounce out into a new phase of expansion. This time too the expansion would also be arrested eventually, and a new contraction would set in, leading to another crunch–bang, and so on. Such a cosmos is cyclic in behaviour, oscillating between a maximum at full expansion and a minimum at the crunch–bang. Each individual cycle has a finite duration, but the history of the universe is infinitely old. Present observations reveal that luminous matter (stars and gas) only consti-tutes about one per cent of the quantity of material required if the cosmological expansion is ever to be reversed to produce a crunch. Nevertheless there may well be vast quantities of inconspicuous mass in the form of black holes, intergalactic gas, weakly interacting subatomic particles or gravitational waves. The idea cannot be ruled out.

The cyclic universe also suffers from the black hole accumulation problem. During the crunch–bang epochs it is hard to see how the compressed matter would avoid falling into more and more black holes with each cycle. There can hardly have been an infinite number of cycles preceding our own, or there would surely be very little matter left. Nearly all the universe would be made of black holes, which is completely incompatible with observation.

Quite apart from all these problems, physicists are reluctant to accept that the universe existed before the big bang for the simple reason that it would mean accepting the worst kind of naked singularity. Given that the Hawking–Penrose theorems predict a big bang singularity somewhere in the universe, the most acceptable place for it to reside is at the beginning of time, where in a sense it instantaneously finishes. If there is no beginning of time, it either means that the singularity is always around, naked, or that a new singularity is likely to form each time the universe collapses, only to last for a short duration and disappear. In that case one could in principle construct a similar naked singularity from a localized mini-collapse, modelling the cosmic collapse. This is the very situation considered in the previous chapter, and regarded as anathema by scientists.

What, then, happened before the big bang? The simple answer is 'nothing', for there was no 'before'. If the big bang singularity is accepted as a complete past temporal boundary of all the physical universe, then time itself only came into existence at the big bang. It is meaningless to talk about a 'before'. In the same way it is meaningless to ask what caused the big bang, for causality implies time; there were no events that preceded the singularity.

The vexed question of whether it is possible, or even meaningful, for time to have a beginning or ending has been debated by philosophers for over two thousand years. Aristotle was of the opinion that the universe is infinitely old, but the Judaeo-Christian tradition, with its central dogma of a creation, seemed in conflict with this. Much later Leibniz speculated that God 'either created nothing at all, or . . . he created the world before any assignable time, that is, . . . the world is eternal', a conclusion which follows from the belief that 'God does nothing without reason and no reason can be given why he did not create the world sooner'. Nevertheless, Leibniz still inclined to the contrary view of a creation, for theological reasons.

These issues were taken up by the eighteenth-century philosopher Emmanuel Kant, in his so-called 'Antinomies of Pure Reason', published in 1781. Regarding the proposal that the universe had no beginning in time, Kant concluded that this would imply that every moment of time had an infinity of preceding moments. Therefore an infinity of successive states or conditions of the world would have occurred. Because Kant believed that an infinite series of states could not be 'completed by successive synthesis' (i.e. infinity can never

167

actually be achieved), he concluded that the idea of an eternal universe was false. On the other hand, if the universe was created a finite time ago, reasoned Kant, then there must have been a time when the universe did not exist. He then argued that nothing can originate from a time when nothing exists, so that 'the world cannot have a beginning'. He thus arrived at a contradiction.

Today we can see the naivety of Kant's reasoning, for the creation of a universe at a finite time in the past does not necessitate the assumption that there was a time when nothing existed. Time itself can be created, an idea that seems to have been anticipated by Saint Augustine who wrote 'the world was made with time and not in time'.

One problem for Kant and others seems to be the dubious logical status of the concept of a first moment. If the universe is not infinitely old, it seems that there must have been a first moment of time. Modern philosophers argue strongly that a first event cannot be of the same kind as other events. Yet many of them miss the fact that a creation 15 billion years ago does not at all imply the necessity of a first event. This seemingly paradoxical assertion can be verified by recalling some of the strange properties of infinite sets discussed in chapter 2. If the universe really did emerge from a singularity, then the singularity cannot itself be considered as belonging to spacetime – it represents, as discussed at length in the preceding chapters, a breakdown of the spacetime concept. If the singularity is not part of spacetime then it is not an event, and did not 'occur' at 'a moment'. But if the singularity is not the first event, what was the earliest moment *after* the singularity?

The question is similar to asking what is the smallest number greater than zero: a millionth, a trillionth? Clearly, any number we like to choose, however small, can always be halved. There is no smallest number. Similarly, there need be no first moment, even though past time is finite. This possibility, which was raised in Tom Stoppard's plays *Jumpers*, alleviates many of the philosophical problems about accepting the finitude of the past.

In the traditional religious picture of the creation, God is responsible for creating the remarkably elaborate cosmic order that we observe around us. We live in such a highly organized cosmos, full of interesting structure and activity, that many people find it impossible to believe it has come about purely by chance and random arrangement. The ultimate in elaborate organization is the human body and mind, but everywhere we look nature seems so well organized that it is easy

168

to envisage the whole universe as the product of intelligent manipulation.

Identifying God with the organizational force is not, of course, an explanation, but a definition. These days, most theologians are prepared to accept that the universe runs itself without the need for continual supervision by a deity. Instead, the laws of nature are able to regulate all natural activity without supernatural assistance. Nevertheless, God is still invoked to set the system going in the first place. In scientific language, God must be manifesting his powers through the naked singularity that marks the big bang creation.

There is certainly no incompatibility between these theological ideas and the scientific version, because the singularity, by definition, transcends the laws of nature. It is the one place in the universe where there is room, even for the most hard-nosed materialist, to admit God. Yet surely a God that is pushed off the very edge of spacetime is a pale shadow of the deity that most people would wish to accept. In this fascinating subject area, where science mingles with religion and philosophy, the urge to push science to its limits is compulsive. Can our, albeit fragmentary, knowledge of singularities reveal anything about the nature of the god who created the universe, to use theological language? Are the properties of naked singularities consistent with the idea of a god who initiates all activity by creating the highly organized structure that we called the universe, and then letting the system run itself according to the laws of nature?

In the previous chapter it was explained how singularities could be either of the organizational or chaotic variety. One can certainly envisage a naked singularity that tosses out ready-made, highly ordered systems. Indeed, it would be possible to believe in a naked singularity that simply vomits forth whole stars and planets, complete with inhabitants! Such absurd speculations are only an expression of the fact that the laws of nature are suspended at a singularity. The more believable image is that of the chaotic singularity, where the breakdown of law leads to complete randomness, so that the emerging material and influences have no in-built organization at all. Any emerging structure is purely accidental and exceedingly improbable.

What evidence do we have from our observations of the primeval universe that the big bang singularity was responsible for creating the high degree of order that we observe today in the universe? When the physical conditions in the big bang are examined it appears that none of the presently observed cosmic organization existed at the beginning. There were no galaxies, no stars or planets, no people, no atoms, and not even atomic nuclei. Any attempt to explain the elaborately arranged and beautifully designed universe that we now inhabit must rest on an examination of the events that have occurred *since* the big bang. The primeval universe was, as far as we can tell, close to total chaos.

The best evidence for the utterly chaotic and disordered beginning is contained in the observations of the cosmic microwave background radiation mentioned earlier in this chapter, this being a relic of the primeval heat. As remarked, the energy spectrum of this radiation is very close to that which is emitted by a body that has come into thermal equilibrium (known as a black body spectrum). At the end of the last century, physicists began to investigate the meaning of thermal equilibrium in the light of the atomic theory of matter. They found that the equilibrium state is the state of maximum atomic disorder, which occurs when the atomic constituents of a body are disarranged in the most haphazard way. The relic of the primeval heat that still bathes the universe carries the unmistakable stamp of atomic chaos in the shape of its energy spectrum. The universe apparently began in total atomic disorder. The way in which cosmic order has arisen out of primeval chaos can be understood in detail by examining the nuclear processes that occurred in the first few minutes after the beginning of the big bang. There is no apparent need for a supernatural organizer – the laws of nature themselves seem capable of generating the present high degree of structure and organization that makes the universe so interesting. A full account of these issues can be found in my book *The Runaway Universe*.

These considerations lead us to believe that, at least in the case of the big bang singularity, the influences that emerged were totally disordered and chaotic. This conclusion may have to be modified in the light of future developments in astronomy and fundamental physics, but it must be admitted that, at the present state of our understanding, science does not support the religious picture of a creator who produced a ready-made cosmic organization. The old idea of a sort of 'package universe', set up in cosmic splendour, does not accord well with the evidence. The organization has emerged slowly and apparently automatically from a fiery start.

This conclusion completes the steady retreat suffered by the concept of an organizational-manipulative deity that began two or three millennia ago. Our ancestors, who invoked supernatural agencies whenever anything inexplicable occurred, gradually learned to replace *ad hoc* supernatural interference with disciplined laws of nature. The world model slowly underwent a metamorphosis from a toy of whimsical gods to an ordered, lawful system – almost a mechanistic device. The domain of the supernatural was gradually eroded as more and more phenomena came to be explained on the basis of scientific principles. By the last century only the creation of man and the creation of the universe remained as legitimate 'areas of operation' for the deity. Today we understand how man, and even (in outline) how life, has arisen on Earth. Now we are also beginning to understand the creation of the universe itself – including the creation of space and time. The organizational-manipulative God has been displaced right back through time, even off the edge of time, banished to a domain beyond the natural world. And we see that in the one place where the natural world meets the supernatural world – the singularity – an organizational God is not really needed either. The cosmos as now ordered has arisen automatically from primeval chaos.

It is truly remarkable that religious adherents have not learned the lesson of history that nature can order its own affairs. The great mistake of theology is to cling to the prehistoric belief in an organizational-manipulative deity. All along they seem to have missed the crucial point that the true splendour of the cosmos is not in the initiation of the organization, but in the laws of nature that nurture and sustain that organization, and operate the cosmic system in an ordered way. What is the virtue in a god, beyond the edge of infinity, who is not part of the stunning beauty contained in nature's mathematical laws? No purpose will be found in the creation by concentrating on issues of causality, for we know that cause and effect are very complicated and subtle when spacetime structure is taken into account. Only a god that transcends spacetime, that is above causality and manipulation, can have any real relevance for the natural activity that blazes all around us.

9 Beyond the infinite

Science acts like an extension of our senses. It allows us to probe into regions of space and time that will for ever remain physically inaccessible to us. Using mathematics combined with local experiments, physicists and astronomers can gradually unravel the secrets of the universe, exploring in their imagination the powerful forces that lie buried in the hearts of quasars, the experiences of ultra-crushed neutronic matter or the weird, space-bending effects inside black holes. They can also turn their attention inwards, to the world within the atom, unlocking a whole universe of the microcosm, populated by exotic and ephemeral wave-particles of many different species, some of which only live for the briefest duration.

The domain of current scientific inquiry extends across a mind-boggling range of magnitudes. Experiments in subatomic accelerators enable physicists to test the laws of nature and examine the structure of matter over lengths as small as a ten-million-billionth of a centimetre and over durations as short as a billion-billion-billionth of a second. At the other extreme, modern astronomy can reach the farthest edges of the cosmos, many billions of light years away. Other physical quantities too span a vast range. The antics of an electron, with a mass of only a billion-billion-billionth of a gram, are as well understood as the solar system with its billion billion billion tonnes of matter.

Year after year, new developments in technology and advances in theoretical understanding extend and consolidate the range of natural phenomena brought within our grasp. Some scientists and philosophers believe that eventually all physical systems will be completely understood and science will cease. The great corpus of human knowledge will encompass all of creation. Others deny that all the laws of physics, let alone a comprehension of their application, will ever be found, however long we toil. Now, with the arrival of spacetime singularities, there are a few physicists who declare that we have encountered the ultimate unknowable anyway, an appendage to reality that will by definition remain for ever beyond the realms of intellectual inquiry.

Can we believe that rational principles are really suspended at a singularity? Does it represent a boundary between the natural and the supernatural – the knowable and the unknowable – or merely between the known and the unknown? There is no unanimous agreement among scientists about these ultimate questions. It could be that the singularity represents the end of the road for science, or else simply a breakdown of the presently conceived laws of nature as new laws, yet to be uncovered, came into play. While we cannot answer these pressing questions at the time of writing, there are at least historical precedents.

Writing in the book *Gravitation* (Freeman, 1973, co-authored with C.W. Misner and K.S. Thorne) under the title of 'Beyond the end of time: gravitational collapse as the greatest crisis in physics of all time' the physicist John Wheeler draws a parallel between the crisis of gravitational collapse and the earlier crisis that afflicted the theory of the atom, which was also threatened with catastrophic collapse. This topic was touched upon briefly at the end of chapter 2, where it was pointed out that the so-called quantum theory of matter, which requires a subatomic particle such as an electron to move according to wave-like rather than particle-like principles, came to the rescue. Physics survived the collapsing atom. Can it also survive the collapse of stars?

Wheeler thinks so. For him, gravitational collapse is a paradox, because on the one hand Einstein's theory of gravity insists that spacetime must come to an end and all of physics must stop, while on the other physics says 'there is no end', for 'physics is by definition that which does go on its eternal way despite all the shadowy changes in the surface appearance of reality'. Wheeler tightens the parallel between the 1911 atomic crisis and today's enigma of gravitational collapse. It was explained in chapter 2 how an electron released near a proton at first sight appears doomed, for it will fall into the proton after a brief duration and shower the surrounding world with an infinite quantity of radiant energy as it disappears into oblivion. This is similar to the infinite compaction of a collapsing star achieved in a finite duration once it can no longer be supported against its own immense weight.

When physics is faced with a crisis of this type there are two routes out. One is to deny the theory that predicts the crisis, and search for a modified theory that closely resembles the existing one in the range of experience that we have so far. This type of patch-up tinkering is the 'small solution' approach. The alternative – the big solution – is to adhere to the spirit of the original theory, but introduce a fundamental new principle that does not alter the theory so much as transcend it.

The atomic crisis of 1911 presented such a challenge to science. The existing theory of James Clerk Maxwell that so beautifully describes the electric and magnetic behaviour of matter predicted most emphatically that the electrons orbiting in atoms would spiral into the nucleus causing the atom to collapse. The reason for this kamikaze behaviour lies with the inexorable tendency for an accelerating electron to radiate electromagnetic waves. Various 'cheap' ways out were proposed to rescue physics. For example, if the simple and fundamental inverse square law of electric attraction between the electron and proton were modified in some way at very short distances, the collapse might be halted.

How can we be sure that Maxwell's theory of electromagnetic forces remains correct down to arbitrarily small sizes? In 1911 experiments had only tested Maxwell's theory down to distances of a few millimetres and up to length of perhaps a few kilometres. Was it reasonable to suppose that the theory of electromagnetism applying to everyday phenomena such as radio transmission should continue to apply right down in the microcosm, over distances as small as a few billionths of a centimetre? When faced by a question of that sort, physicists can only make a judgment using non-scientific criteria. Maxwell's theory is agreed by all to be mathematically and physically a most elegant and succinct description of electromagnetic forces, and by some to be the most beautiful intellectual construction of the nineteenth century. Anyone who has studied Maxwell's electrodynamics cannot fail to be struck by the internal symmetry of the mathematics and the compact arrangement of the interplay between electric and magnetic effects. Such a simple and satisfying theory of nature carries a compulsion which is quite outside the narrow criteria that scientists can use to test their theories in the laboratory. This is why Maxwell's electrodynamics is held in such high esteem for aesthetic as well as physical reasons. To tinker with it in a crude attempt to avert atomic collapse is almost blasphemy. It is like entering the Taj Mahal to find it supported on concrete blocks. In the end, not only did Maxwell's theory survive the atomic crisis, but today it has been checked over distances that range from a hundred-millionth of the size of the atom up to the dimensions of a galaxy.

Another cheap way out of the atomic crisis was to deny that an electron in orbit about an atomic nucleus does emit electromagnetic radiation, even though laboratory experience had shown that electromagnetic waves do radiate from electrical systems of 'everyday' dimensions. This attempt comes into head–on conflict with Einstein's special theory of relativity, which in 1911 was still relatively new. The problem here is that when an electron moves from one side of an atom to another, the electric field that it carries must move with it. Now this field can be detected (in principle at least) far away from the electron. If the field were to readjust itself instantaneously to the new location of the charged electron, then we would have at hand a means of signalling. For example, a simple code arrangement could be set up whereby if the electron is stopped in its orbital path by the signaller, then the abrupt cessation in the motion of its electric field recorded some distance away means 'yes'. Failure to stop the electron means 'no'.

The conflict with the theory of relativity is that the transmission of information (even of such limited scope) is absolutely forbidden to take place faster than the speed of light. It follows that the electric field of the electron cannot, after all, readjust itself precisely instantaneously at all distances from the electron. The nearby field can adjust rapidly, but the more distant regions of the field have to wait until at least the light-travel time from the electron before they can change. The edges of the field are therefore sluggish to respond and lag behind the field that is near to the electron. When the electron changes its motion suddenly, such as when it orbits around the atomic nucleus, the far field trails behind it in its motion. Because the nearby field changes more rapidly than the far field, the field becomes distorted in shape by the electron's accelerated motion. This disortion or kink travels outwards with the speed of light as the more distant regions of the field 'catch up' with the changes in the nearby field. But the outward-travelling kink carries energy and momentum with it and is, in fact, precisely what we call electromagnetic radiation. The requirement that instantaneous signalling be impossible leads inevitably to the flow of radiant energy away from the accelerated electron. To prevent atomic collapse by denying that the electron radiates amounts to accepting faster-than-light signalling and with it the attendant causal paradoxes mentioned in chapter 3. We have to choose between abandoning causality (and allowing signals to travel into the past) or finding a new principle that transcends the old concepts.

In the event, the new principle was discovered by Niels Bohr and elaborated in the 1920s by Erwin Schrödinger and Werner Heisenberg. This so-called quantum theory embodied both the structure of Maxwell's electrodynamics and the principles of causality, but embedded them in a new conceptual framework that did not allow atoms to collapse to a single point amid an infinite quantity of radiation. Instead the curious wavelike behaviour of subatomic matter was discovered and seen to provide a cushion of energy that prevents the electron from being confined too closely around the atomic nucleus. It took a change in the fundamental structure of physics to solve the atomic crisis.

Wheeler regards today's controversy about gravitational collapse and the appearance of spacetime singularities as a re-run of these earlier debates about the atom. Einstein's general theory of relativity is regarded by many as the supreme intellectual achievement of the human species; certainly it surpasses Maxwell's electromagnetic theory in elegance, economy and scope. It builds the world out of pure geometry and arranges it according to mathematical laws that could scarcely be simpler or more powerful. Few scientists would deny that its aesthetic appeal is the most persuasive evidence in its favour.

Yet Einstein's theory leads irresistibly to a singularity, to unbounded gravitational collapse. It is frequently proposed that the theory should be abandoned in the face of this absurdity. Arguments are advanced that Einstein's relativity has only been tested over distances varying between a few metres to possibly a few billion light years. How can we be confident that it continues to apply in the microcosm? Is it so surprising if the theory fails, say, within the atom? Perhaps when a collapsing star shrinks to a billionth of a centimetre in size, the gravitational inverse square law fails?

The aesthetic arguments that can be used to defend Maxwell's theory in the face of atomic collapse can be deployed even more convincingly in favour of Einstein's theory. Tinkering with this great edifice of descriptive and predictive power in order to alleviate the singularity crisis seems like a 'cop-out'. It was not the way out in 1911, and it would be surprising if it were the solution today.

There is also a close parallel to the attempt to solve the atomic collapse problem by abandoning causality. At first sight the idea seems very attractive. In the earlier chapters it was explained how a burnt-out star cannot summon up the necessary heat pressure to resist progressive shrinkage. The cores of many stars become so compact that their atoms are crushed into neutrons, and they end their days as neutron stars only a few kilometres across. It was mentioned in chapter 5 that if a neutron star were to contain more than a few solar masses, then even the neutronic matter would not be stiff enough to resist further, and catastrophic, collapse.

It is a familiar experience that the more a material is compressed, the stiffer it becomes. A neutron star is already a billion times stiffer than steel, and yet still gravity overwhelms it. Nevertheless, could it not be that, in the end, as the material is further compressed during the collapse, the stiffness of the supercrushed neutrons will rise enough to prevent further shrinkage? Given that we know nothing at all about the behaviour of matter under such extreme conditions, what grounds are there for supposing that gravity will always be able to overcome the rigidity of matter, however far it is compressed?

It is true that we are unlikely ever to know fully the properties of neutronic matter, let alone still denser states, yet there is a general and powerful argument against invoking superstiff matter as another cheap way of avoiding singularities. The considerations involve, rather surprisingly, the somewhat mundane phenomenon of sound. Musicians are familiar with the fact that a stiff, taut guitar string plays a higher note than a floppy string. Physically this is because the vibration frequency of the string is higher. If the vibration is regarded as a standing wave, i.e. a wave trapped between the fixed ends of the string, then one can infer that the high frequency is due to the high speed of wave propagation on the taut string.

The relation between stiffness and wave speed is easily demonstrated. A long piece of string or elastic, if plucked at one end, will send the disturbance shooting down its length. The more taut the string, the faster the kink travels. Simple experiments with solid materials also reveal that sound waves travel through them at varying speeds

177

depending on their stiffness. It is a property that is simple to understand. Sound waves are, after all, vibrations of the material. If a rod of metal such as steel is tapped at one end, the material compresses slightly beneath the blow. Being elastic the steel tries to restore itself to its natural density, and in so doing propagates the compression to the neighbouring material. This too tries to relax back, and nudges the material further down, and so on. The effect of the elasticity of the steel is to cause the compression of the blow to propagate rapidly along the length of the rod. This is sound: similar compression waves propagate through the air to our ears. If the material is very stiff, then the elastic forces are strong and the compression is fought vigorously and rapidly. (Taut elastic snaps back faster than floppy elastic.) Thus, in a stiff material the sound wave travels very fast.

When it comes to neutronic matter, the elastic properties are unlike anything known on Earth. A neutron star is so stiff that the speed of sound is an appreciable fraction of the speed of light. It is a simple matter to investigate the elastic properties of a still stiffer material using mathematical modelling, and the result deals a death blow to hopes of averting catastrophic gravitational collapse. It turns out that any material that is stiff enough to support a shrunken star of several solar masses will have to propagate sound waves through its interior at a speed that exceeds that of light waves. Picturesquely speaking, one could shout a message faster than waving it. This is strictly forbidden by the theory of relativity; no information can travel faster than light, including the spoken word. To avert total collapse by appealing to ultrastiff matter is to abandon causality, in the same way that the aversion of atomic collapse by invoking the absence of radiation violated causality. If we use the lessons of 1911 to guide us, this is not the way forward.

What, then, of the quantum theory, so successful in stabilizing the atom? Unfortunately this question is still open because there is as yet no viable quantum theory of the gravitational field. As explained at the end of chapter 2, all quantum theories of the forces of nature run into problems with infinity, and only the so-called renormalizable theories, in which the infinities can be swept under the carpet by some simple mathematical tricks, have any real value when it comes to prediction. The quantum theory of gravity is uncontrollably polluted with infinite quantities that do not in any way seem to be related to the infinities of the spacetime singularity. It is therefore not possible to perform the equivalent of Niels Bohr's atomic computation to see if quantum effects can save the universe from the singularity.

In spite of the paucity of concrete calculations, there has been plenty of speculation about the likely effects of quantum gravity. In an atom the quantum 'cushion' energy that prevents the electron from spiralling in towards the nucleus operates effectively at a distance of a few billionths of a centimetre. An electron, however, is attracted to the nucleus (electrically) by a force that is almost infinitesimal by the standards of gravitational collapse. If a star is shrunk down to the size of an atom, the gravitational force pulling an electron inwards is a billion billion times greater, and quantum effects are utterly negligible. Only when the star has shrunk to a size twenty five powers of ten smaller than an atom will quantum theory begin to play a role. This means that the star is a mere million–billion–billion–billionth of a centimetre across!

At such unimaginably small lengths, profound modifications to the structure of space and time can be expected. It has been conjectured that the quantum disturbance will be so severe that even the topology of spacetime will alter. Instead of a 'bumpy sheet', it will display a foam-like structure, full of worm-holes and bridges. The concept of a continuous space and time, entrenched for centuries, seems to be threatened when quantum effects are taken into account. Nobody has the slightest idea of what would happen to the collapsing star when it gets among the spacetime foam. Here we enter an area of physics for which there is very little guidance from established principles. It may be that a singularity will still occur. Nobody knows.

If, as Wheeler suggests, we have the courage to stay with Einstein's theory, and if we believe that physics must continue in spite of gravitational collapse, then we are inevitably led to assume that radical new physics must take over at some stage before infinity occurs. Probably the whole concept of space and time will have to be abandoned in this new domain, so that the singularity will have to be regarded no longer as an end to existence, but merely as an end of spacetime. Wheeler likens the situation to the investigation of elastic media. Both spacetime and elastic can be bent and distorted. How can we find out about the elastic forces in a material? In particular, how can the elastic limit, when the fabric snaps, be predicted? The study of elasticity is full of beautiful mathematical theory and impressive predictions covering a wide range of phenomena. The properties of elastic solids, liquids and gases form part of everyday experience and are familiar to every engineer. Yet a lifetime of engineering experiments will contribute nothing to the understanding of the origin and limits of elasticity. To find out why rubber or steel can be stretched and compressed one has to study the atoms and molecules from which these substances are made. We have to experiment within the microcosm to explain the forces that bind the molecules together and provide the bulk material with its elasticity. A study of elastic media can never explain atoms, but atoms can explain the elasticity of macroscopic materials.

The progress of science has traditionally been along a well-defined route that seeks to explain the gross properties of nature in terms of the behaviour of their smaller constituents. By examining matter at an ever-finer scale of distances, most of the macroscopic properties are now understood. It seems reasonable that the same basic route should also apply to spacetime: further and further analysis into smaller and smaller units should eventually uncover the 'atoms' out of which space and time are built.

It would have appeared nonsensical to Newton, and certainly to Leibniz, to talk about building space and time out of microscopic units after the fashion of solid material. In this century, with Einstein's theory of gravity describing a sort of 'elastic' distortion of spacetime, the idea does not appear so ridiculous. The concepts of space and time, however, are so deep-rooted in our culture that they are not abandoned easily. With no conceivable experiment to provide guidance, only mathematics and aesthetics can suggest what entities may lie buried within the spacetime fabric. Several valiant attempts have been made to build up the essential properties of space and time out of abstract mathematical objects that are otherwise unknown in physics. Ultimately one might hope that even topological properties such as the dimensionality of space and time (three and one respectively) would emerge naturally from such a treatment. Unfortunately the mathematical problems are formidable and these pregeometrical theories are still in their infancy.

When gravity becomes too intense for geometry to survive, space and time come apart, and one reaches the equivalent of the elastic limit of a solid, where the material snaps under the strain. If there is a pregeometrical substructure it will be here, on the edge of infinity, that it will be exposed. If there really do exist naked singularities then there is a chance that we can witness the substructure at work, for the singularity will be able to send information into the surrounding space. Until that happens, only speculation can respond to the burning questions about the fate of collapsing stars, or perhaps of the entire universe.

As a collapsing star shrinks down and down it will eventually tear open spacetime and enter the world of pregeometry. We do not know what may be encountered in that world, or what properties it will have. We do not even know the entities of which it is composed. If physics does continue into this region beyond spacetime then the star would seem to be faced with two alternative fates. Either it disappears for ever from spacetime, and joins the denizens of this mysterious pregeometrical substructure locked, as it were, inside spacetime, or it will find that the pregeometrical region is only a thin barrier to some as yet unknown further region, perhaps another spacetime – a parallel universe like those mentioned in chapters 4 and 6.

This raises the issue of the creation event – the big bang – and what may have lain before it. If a collapsing star in some sense survives, albeit in a completely mutated form, we have to conclude that the universe did, after all, have some existence before the big bang. Did the universe that we now perceive erupt from this mysterious pre-geometrical world 15 billion years ago? If so, was the big bang only a transient phase between successive cycles of the universe? Some of the problems encountered by such a theory have been discussed already in chapter 8, but if the pregeometrical world really exists then all manner of physical laws could be suspended there.

Indeed, John Wheeler has argued that the universe is a cyclic mechanism, alternately expanding to a huge size and then collapsing to an ultradense, microscopic state of pregeometry, from which it emerges in a big bang with all physical systems 'reprocessed'. In this 'new deal' cosmos, each cycle of expansion and contraction begins with fresh subatomic particles, new heat, new motion and maybe even new laws of physics. Possibly the features of the emergent universe are chosen purely at random, like the throw of a cosmic die. One can envisage worlds that are totally alien to our experience, in which even the nature of matter is quite different. Whether or not life of any sort could exist in these other cycles is a much-considered question. At present, it seems that life as we know it is rather delicately balanced, so that any major restructuring of the physical world that we experience now would have dire consequences for biology.

The route to the naked singularity has been a long and challenging one. On the way we have encountered exotic and bizarre systems, strange physical concepts and stunning mathematical notions. The journey began with the story of gravity and its all-embracing power. The humble inverse square law, learned by every school child, turned out to threaten matter in a way that could scarcely be over-emphasized. We began to see the universe as a battleground between gravity and the other forces of nature, the former trying to crush matter into non-existence, the latter resisting, more or less hopelessly. The fact that we are alive to witness this battleground is testimony to the fact that gravity must fight hard for every victim, but eventually it seems assured of success. Most of the cosmos is made of stars, and stars pay dearly for their temporary reprieve. Every day it costs the sun a million billion billion dollars worth of energy to escape near-instantaneous implosion into nothing of which we know. Eventually the credit runs out and the debt of annihilation is paid.

We saw how the demise of a star can be a complicated affair, involving titanic explosions and turbulent activity in the centre. The collapsing core should, if all is well, rapidly become enveloped in an event horizon, the core itself shrivelling in a few millionths of a second into a black hole from which it can never return. We followed the fate of the core as its size dwindles ever more rapidly and we studied the almost grotesque distortions that this gravitational violence imprints upon the space and time around it. The singularity emerged as the ultimate limit of physics – the end-point of all known laws. It appears inexorably, as the inescapable outcome of the formation of a trapped surface that can bend light back on itself and twist space into time.

We encountered two opposing attitudes to the singularity. The first regards it as the edge of existence, the infinite that bounds the finite – a non-place where the natural world ceases. The alternative viewpoint, so eloquently expounded by John Wheeler, views the singularity as the threshold where space and time are transcended, but the physical world, in some suitably encompassing sense, survives. As what, nobody knows. Only a naked singularity can answer that question.

Many of the issues that have been treated in this book seem to be closer to philosophy and religion than the sort of science that is so often associated with our daily experience. Practitioners of these subjects – cosmology, gravitation and relativity, spacetime structure, and so on – are frequently castigated for their irrelevance to the 'true path' of science, i.e. the service of man through technology.

The history of science in this respect shows a curious evolution. In the most primitive societies, what we would normally call science was completely absent. Yet the development of symbolic communication (pictures, writing, sculpture, etc.) opened up the possibility of organizing knowledge and experience for the benefit of the wider community. Although the concept of an experiment is relatively recent (post–Galileo), early man must have learned by trial and error all the rudiments of an ordered society: how to make weapons and agricultural implements, how to build houses and then cities, how to irrigate crops, and so on. Slowly but surely, in a random sort of way, primitive cultures gained control over their environment and recorded the techniques for posterity. Science, even as the hit and miss affair that it was, nevertheless directly concerned the technological needs of the society.

Later, in what is nebulously called Ancient Greece, we see science – and also the rapidly developing subject of mathematics – running along a parallel course. To be sure, the demands of technology, especially military advantage, still required an intellectual input, but most of the truly monumental wisdom passed down to us from classical times stemmed not from the battlefield or the workshop, but from the philosophers and theologians. The technique of inquiry was, naturally, quite different from today. Disputes about such issues as the existence of atoms, the laws of motion or the organization of the cosmos were not resolved by experiment or carefully managed observation, but by appeal to theological principles based on blind dogma.

The tradition of 'science' as a province of theology became so firmly established in the centuries that followed that during the European Dark Ages, the Christian priesthood and the world of Islam were almost alone responsible for the perpetuation of knowledge in the Western world. By the time of the European Renaissance, the Church had such a hold on 'scientific' matters that dissenters were frequently threatened with a choice between death or doctrinaire compliance. In many ways it seems as though science was solely in the service of religion, and issues such as the structure of matter, the nature of space and time or the origin and end of the universe were regarded as legitimate, even worthy, subjects of study, for the betterment of Christianity.

Historians will dispute the root causes for the disintegration of the theological stranglehold, but a contributory factor must have been the growth of world trade and the discoveries in the New World. Increasingly, as nations became more complex, so technology was needed to gain advantage over competitors in trade, colonization and warfare. Sophisticated navigational devices were needed, modern mobile weapons were in demand and the burgeoning trade with the East also brought fresh problems of transportation, storage and distribution. Science and mathematics were intruding more and more into the area of practical gadgetry, from the telescope to the spinning jenny.

The industrial revolution that followed the liberation of science from its theological yoke sent repercussions down the centuries which still reverberate today. For it is technology, rather than philosophy, that once more demands unswerving attention from the scientist. The use of science for profit or military advantage now goes under the name of 'applied science'. The pursuit of knowledge, either for its own sake or in the service of more abstract ideals such as finding a purpose for our existence, has come to be known as 'pure science'. Of course, the distinction is an artificial one, and year by year various topics make the transition from pure to applied, and occasionally the reverse.

The importance of this distinction between pure and applied science lies not so much in its epistemological significance as in the practical organization of the subject. Scientific research (in the widest sense) is now a major industry occupying a significant fraction of the world's manpower and resources. Investment in education and research now outstrips any other area of expenditure, even in these days of spiralling arms development. The knowledge business is big.

How should this vast investment be administered? Who should decide what research topics are to receive priority or what academic subject courses should be funded at colleges and universities?

In practice, there are two major forces that operate to influence the distribution of available resources and investment. On the one hand are the demands of the community as a whole, which is asked to allocate part of the common wealth to ventures that are frequently not understood or are even regarded with apprehension by many. On the other hand, the individual scientist will have his or her own motives for pursuing a certain line of research. These might include innate curiosity, bandwaggoning, financial gain, social or ethical considerations, or just plain ability – we do what we are good at.

184

Experience has shown that a compromise must be adopted between academic freedom and concrete return on investment (i.e., 'applications'). A major research effort might be mounted to solve, say, the problem of controlling nuclear fusion as a contribution to alternative energy strategy. Given that controlled fusion is a desirable end (which some people dispute), the obvious way to achieve it would be to assemble a large number of 'experts', institute a crash training programme for young graduates, and pour considerable sums of cash into facilities and experimental equipment. It may then be hoped that after a not unreasonable duration (fifteen years?) the problems will be overcome and the construction of commercial power stations will proceed. This scenario is, in fact, close to reality. Both the US and USSR currently have substantial research programmes into controlled fusion, while the European effort is concentrated at the so-called JET (Joint European Torus) facility at Culham in Berkshire.

Does historical precedent encourage the use of 'head on' tactics for solving outstanding technological problems? Unfortunately, the progress of science owes more to accidental discoveries in seemingly useless (in the profit sense) pursuits than it does to goal–oriented research. While this may not be true for essentially cosmetic gadgetry (inventing a brighter light bulb, a stickier glue, a greener green paint, a less–polluting engine . . .) it seems to be the case for the truly fundamental scientific discoveries.

A few examples will suffice to make the point. Maxwell's mathematical investigation of the relation between electric and magnetic forces must be considered to belong to the realm of pure science. It was undertaken for the sole purpose of advancing the understanding of Maxwell in particular and humanity in general about the nature of electromagnetism and the mathematical principles that underpin it. The outcome of this epochal work was the (mathematical) discovery of electromagnetic waves, with their eventual laboratory production and detection by Heinrich Hertz some years later. The fact that we now inhabit a society strung together by electromagnetic communication of one sort or another stems not from any commercial, military or administrative edict for a crash research programme into telecommunications, but from one quiet Scottish gentleman's nagging suspicion that the existing theory of electricity and magnetism was slightly wrong.

Another historical episode close to us all was the development of flight. A clear commercial pressure existed at the turn of the century for heavier than air flying machines, and much 'head on' research took place into understanding the mechanism of bird flight. Intrepid aviators flapped their way into historical oblivion, while the really significant discovery was occurring elsewhere with the development of the petrol engine, which alone could provide the missing motive power.

Our age is beset with social problems that are being tackled 'head on'. The search for a cure for cancer consumes an outrageously disproportionate amount of cash for the limited progress made. Yet almost certainly any cure would not come from a direct investigation of the disease, but rather through an advance in fundamental biochemistry. And the greatest contribution to solving the world's food shortage lies not in the obvious – finding ways to grow more food – but in efficient contraceptives.

In a society as complex as ours, the 'head on' approach to problems is frequently the most wasteful. The controlled nuclear fusion projects, for instance, have now dragged on uninspiringly for over a generation. Often the solution pops up out of some totally unexpected area that had no obvious relation to the issue at hand. Stories such as Antoine Henri Becquerel's accidental discovery of radioactivity (when he noticed a mysteriously fogged photographic plate) or Alexander Fleming's luck in spotting the inhibitory effect of some mould on a bacterium culture that led him to penicillin, reinforce the case for the unshackled scientist, following his own intuition rather than the orders of his paymaster.

It is the recognition that the pure science of today may become the applied science of tomorrow that persuades cost-conscious governments to continue to support fundamental research for its own sake. Fitting squarely into this category come astronomy and astrophysics, cosmology, and to a large extent the theories of gravity, space and time, and subatomic matter. A hundred years ago, all these topics were the preoccupation of individual scientists, or small groups working in laboratories that were little more than sheds, and with budgets that represented a negligible fraction of the resources of their institution. Today, most 'pure' scientists work in large laboratory teams, usually part of a university, and annual national budgets run into hundreds of millions of dollars. A new telescope, or satellite, or the next generation of atom smashers, may individually take many millions from dwindling grants. No longer is it possible for great discoveries to be made with sealing wax and wire in the attic. Science has become institutionalized and integrated into the socio-economic framework.

Sometimes pure science research programmes are regarded by the rich nations as prestige projects. Nobel prizes represent a sort of national intellectual virility symbol, and much research effort is duplicated in the race to obtain them. Some may view this as healthy competition, but many will see it as a wasteful game.

There is another side to pure science as well as prestige and unexpected spin-off, and that may be called, loosely speaking, human interest. The truly fundamental issues – the nature of the universe and the place of mankind in it – will always be somewhere in the back of everybody's mind. Science may no longer be the servant of religion, but it still contributes immeasurably to the wider philosophical perspectives that lie above the clamour of technological society. In chapter 1 we examined the historical origins of the cosmic connection and saw how close to human thought astrology and astronomy had been for many centuries. In this age, despite the distractions of material things, the cosmic connection is still there. The Earth, now more than ever an insignificant speck in the vastness of space, can be seen as only a tiny component in the great scheme.

For many people, the discoveries of modern astronomy will remain for ever remote and inconsequential. What have pulsars or quasars to do with the problems of family life or the day-to-day toil? Yet there is more to life than merely getting through it, and the history of mankind is full of the struggle of our species to establish its precedence through technology and to expand its stock of knowledge by inquiry. Indeed, it is only in this century that so many people are questioning the value of fundamental research. Our ancestors would have regarded deliberation on the cosmic arrangement as a perfectly natural enterprise.

What, then, of the future? If pure science does not grind to a halt through lack of support or the sheer cost in operating expensive laboratories, can we expect the next hundred years to be as productive as the last?

In this book I have dwelt at length on some of the most fundamental and challenging topics in science today. Many of the things discussed, especially the nature of black holes, quasars, the early stages of the big bang, and the ultimate consequences of gravitational collapse, are only vaguely perceived as yet. Twenty years ago none of these subjects existed. Nevertheless, they seem to present science with a crisis that is unparalleled since the turn of the century. Two outcomes are possible: either we are at the end of the road or we are on the threshold of a new revolution in physics.

Scientific progress has become so commonplace that many people accept it as a law of nature. It seems to be almost universally assumed that however enigmatic an aspect of nature may appear, sooner or later science will come up with the answer. Certainly in the last couple of centuries scientific advance has been so spectacular that it is easy to take it for granted that the pace of progress and discovery will continue to grow.

In spite of this, there is no basic principle that requires nature obligingly to drop clues into our laps whenever we get stuck on a problem. The physicist Eugene Wigner has spoken of the 'unreasonable effectiveness of mathematics in the natural sciences'. We are so used to the success of simple mathematical modelling that we scarcely appreciate how stunningly remarkable it is. Einstein's equations can be condensed into the disarmingly simple requirement that the gravitational antics of matter should always be regulated so that two mathematical quantities remain for ever equal. This one equation can be used to derive all of gravitational physics. A similar, equally condensed equation, describes the electromagnetic force. Why does nature let us write down most of her secrets on a single sheet of paper?

It is always amazing that the universe, being such a complex place, is nevertheless ruled by simple, mathematical principles. Sometimes this is expressed by saying that 'God is a mathematician', while others argue that mathematics is specifically invented to describe the world about us, so it is no surprise if that same world turns out to comply with especially simple mathematical expressions. (The reader may well feel, after glancing at a textbook on electrodynamics or relativity, that the word 'simple' is being stretched in its meaning here. That is not so. Modern mathematics may be esoteric and conceptually very hard, but it can at the same time remain elegantly uncomplicated.) Whether mathematical simplicity is God's affair or ours, the fact remains that this feature more than any other remains the mainspring of progress in the physical sciences.

What we have to face is the prospect that the stock of mathematically simple systems in the natural world may be running out. It could be that the physics beyond gravitational collapse – Wheeler's 'pre-geometry' or whatever else it might be – will always lie beyond our mathematical capability to describe, either because the mathematics itself is not advanced enough, or because there simply does not exist any mathematical description of this situation. In the biological sciences one is faced with a bewildering array of complex life forms and processes, very little of which has succumbed to mathematical modelling. Randomness and probability play a large part in both subatomic physics and biology, yet mathematics still fails to cope adequately with the latter. It seems a hopeless task to expect to get an equation that will predict a dog, or contain terms that can be identified with the digestive system of a wombat. Perhaps gravitational collapse is like this – a zoo of possibilities, with no logical reason for one rather than the other, and no method of detailed prediction.

If the singularity leads us to this sort of muddle then it will indeed represent the end of the road for physics as an exact science. On the other hand, the deeper we probe nature, the simpler it seems to become, so perhaps it is not unreasonable to hope that buried in the recesses of the spacetime disruption – on the edge of infinity – we shall uncover a whole new era of physics, more elegant and more fundamental than we have already. This is Wheeler's expectation too. He writes: 'Some day a door will surely open and expose the glittering central mechanism of the world in its beauty and simplicity. Towards the arrival of that day no development holds out more hope than the paradox of gravitational collapse.'

The route to this new physics of the future lies beyond the edge of infinity. What it will tell us we can only conjecture. Certainly it will show that space and time are only approximate constructs; it will reveal how the quantum world, how space and time, and how matter are all put together and relate to each other. It will enable us to go beyond spacetime in our analysis, to a domain that as yet seems to belong elsewhere than in the physical universe. The new physics will answer the question of where a collapsing star goes and what comes out of a naked singularity. It may even tell us where the universe came from and how it will end.

All that will be ours if we can meet the challenge of the singularity.

Index

ABOUT THE AUTHOR

Paul Davies is Professor of Theoretical Physics in the University of Newcastle-upon-Tyne and was formerly a lecturer in applied mathematics at King's College, University of London. He writes internationally for magazines and journals, including *Nature, New Scientist, The Economist* and *The Sciences,* and he frequently contributes to science broadcasts. He is the author of *The Physics of Time Asymmetry; Space and Time in the Modern Universe; The Runaway Universe; The Forces of Nature; The Search for Gravity Waves;* and *Other Worlds.*